D#217706

3/18/86 PА

Under Newton's Shadow

Astronomical Practices in the Seventeenth Century

Under Newton's Shadow

Astronomical Practices in the Seventeenth Century

Lesley Murdin

The Open University

Adam Hilger Ltd
Bristol and Boston

British Library Cataloguing in Publication Data

Murdin, Lesley
    Under Newton's shadow.
    1. Astronomy—Great Britain—History
    I. Title
    520′.941   QB33.G7

    ISBN 0-85274-456-0

Consultant Editor: **Professor A J Meadows**
Department of Astronomy, University of Leicester

Published by Adam Hilger Ltd
Techno House, Redcliffe Way, Bristol BS1 6NX, England
PO Box 230, Accord, MA 02018, USA

Phototypeset in 11/12pt Garamond by Quadraset Ltd, Midsomer Norton, Bath
Printed in Great Britain by J W Arrowsmith Ltd, Bristol

Contents

Acknowledgments   vi

Preface   vii

1   Astronomy   1
2   Astronomers   7
3   Preparation for astronomy   33
4   Personal lives   53
5   Money matters   73
6   Communication   90
7   The tools of the trade   108
8   The astronomer's image   131

Bibliography   147

Notes on manuscript sources   148

Index   150

Acknowledgments

I am grateful to the following for permission to use material from the archives and to reproduce illustrations:
 The Library of the California Institute of Technology
 The Royal Greenwich Observatory, Herstmonceux
 The Royal Society of London.

I should also like to thank the following individuals for generous help:
 David Clark for all his work on Stephen Gray
 David Calvert for help with illustrations
 Janet Dudley for assistance with research
 Joy Hamblyn and Patricia Pinn for secretarial work
 Paul Murdin for constructive criticism of the manuscript.

Preface

The nucleus of this book was some research into the life and work of the amateur astronomer, Stephen Gray. David H Clark, then at the Royal Greenwich Observatory, was interested in the sunspot activity that took place during and after what has become known as the 'Maunder Minimum'. His interest and inspiration brought to notice the work of Stephen Gray whose achievements in both astronomy and electricity seemed to have received very little recognition. Although Gray moved for a long time on the fringes of the Royal Society and was eventually made a Fellow in 1732, the main forum for his activities was outside the Royal Society altogether. In fact, some of his best astronomical work was refused for publication in the Society's journal, the *Philosophical Transactions*. Through research into his letters, and later into other aspects of his life, David and I were gradually able to put together some pictures of the practice of astronomy in the time of Newton.

One of the most fascinating aspects of Gray's life is the apparent improbability of his becoming interested in astronomy and physics at all. He worked as a dyer and constantly complained of the lack of money, books and time for science. Yet, in spite of all the difficulties, he continued to work at his research until the day before his death. The curiosity that held him for so long must have been a powerful force. What we discovered about Gray's life raised many questions about how he became involved in astronomy and how he managed to put his interest into practice. Astronomical observations were being made at this time by numbers of astronomers who, like Gray, were near the beginning of the tradition of observational astronomy in this country. The questions raised about Gray could usefully be extended to other members of a generation that was particularly active in astronomy at all levels, even though overshadowed (as far as posterity was concerned) by the genius of Newton.

Gray's case made clear that looking at the astronomers of the late seventeenth to early eighteenth centuries through the activities of Newton and the Royal Society would mean omitting a large part of what was happening. To discover how astronomy was practised by the lesser men in the time of Newton required an examination of wider correspondence and records.

An excellent centre for such work is provided by the Flamsteed Papers at the Royal Greenwich Observatory. John Flamsteed was the first Astronomer Royal at the old Greenwich Observatory. By the end of his long life he had quarrelled with most of the leading astronomers of his day, particularly Newton and his colleagues at the Royal Society. Yet at the same time he had a large network of loyal assistants and correspondents, of whom Stephen Gray was one, scattered throughout the country. From a study of Flamsteed's correspondence has emerged a picture of the less well known predecessors of modern astronomers.

This book is an attempt to see why and how so many men from different social backgrounds chose to work at astronomy in the time of Newton. My thanks are due to David Clark for the use of his work and for his encouragement in undertaking the writing. I should also like to thank the Archivist at the Royal Greenwich Observatory, Miss Janet Dudley, for her help and encouragement.

1

Astronomy

Isaac Newton, standing, as he himself put it, 'on the shoulders of giants', cast a long shadow. Others working in astronomy during his lifetime have perhaps received less attention from posterity than they would if they had lived in another age. One of the lesser men who interacted in a small way with Isaac Newton was Stephen Gray.

Gray was the owner of a small dyeing business in Canterbury. Careful research has not been able to discover any evidence that he had a formal education. Yet he became a Fellow of the Royal Society, and at the very end of his life, after working on questions in both astronomy and electricity, he was one of the first to be awarded the Society's Copley Medal. Soon after, he died in a charitable home for respectable but impoverished old men. Gray's story, his embattled relationships with his famous contemporaries, opens up many questions about the practice of astronomy in Newton's time. What sort of men found astronomy so fascinating that even without obvious academic connections they learnt what they needed, found money or patronage, acquired instruments, fought their way through the controversies of the day and eventually made some contribution of lasting value to astronomical knowledge? These kinds of questions can be fruitful in leading to a better understanding of a period in which the central figure is so dramatic that he has always received a great deal of the available attention.

Newton lived from 1642 to 1727. In England during those years the science of astronomy was developing a structure and sociology of its own. In earlier times the word *astronomy* had been used both for the study of the motions, compositions and relations of celestial bodies and also for the use

1

of those studies in divination and prediction. In other words, 'astronomy' had included *astrology*.

By the seventeenth century, the work of observers like Galileo and Tycho Brahe and of theorists like Copernicus and Kepler had shown that exciting advances might be made in understanding the heavenly bodies. The invention of the telescope at the beginning of the century meant that observation itself could be more accurate than ever before, continuously producing new data on which new theories could be based.

Outside events, too, particularly the Civil War in the middle of the seventeenth century, changed attitudes to what astronomy was and what it might do. Christopher Hill (1965), among others, has emphasised the practical, pragmatic tendency of Protestantism, and especially the more puritanical shades of Protestantism. After the Revolution there was little chance of a large-scale return to the magical, mystical view of the world that had been favoured by Catholicism. Moreover, the suspicion and fear caused by a Civil War led to a general desire among astronomers—as among the rest of the population—to avoid the dangers of speculation and conjecture, keeping to *the observation of facts*, as the increasingly influential Chancellor, Bacon, had urged at the beginning of the century.

Astrological predictions and divination had helped to keep the population in a high state of anticipatory tension during the Civil War. Between George Wharton and William Lilly had flown the main volleys in the astrologers' own war in which Royalist and Parliamentarian astrologers had produced predictions favourable each to their own side, and had forecast doom to the enemy.

Since astrology allowed itself to meddle in politics, the most successful astrologers of the period were those who had the political and journalistic skill to provide for any turn of events. The most consistent in the tenor of his predictions was probably George Wharton, who continuously supported the Royalist cause and was thrown into prison during the Commonwealth period. William Lilly was probably the most popular. He supported the Parliamentarians so wholeheartedly that passages from his almanacs were read aloud to encourage the troops into battle. He was such an asset to his side that the Royalists offered him £50 to defect and then tried to kidnap him (Parker 1975). Not surprisingly, he completely failed to predict the Restoration, but claimed that events 'above nature' had led to the return of the monarchy, which, therefore, could not have been predicted by the 'natural science' of astrology. After the death of Oliver Cromwell, Lilly showed sufficient astuteness to write ambiguously flattering remarks that could apply to any victor. With such skilful manoeuvring he survived until natural death in 1681.

During those heady days when astrology was at its most influential, astrologers like Lilly felt no need to work at making accurate astronomical observations. Lilly's own background was medical, and although he wrote a book called *Christian Astrology* in which he attempted to set out clearly the principles of astrology, he would have had little in common with experimental scientists. After the Restoration, however, the new spirit of caution and adherence to observation led to some attempts to establish a firm factual basis for believing in the influence of the planets. Since the nature of this influence was bound to be difficult to investigate, astrologers like John Gadbury set out instead to collect data to demonstrate the workings of planetary influence in practice.

For thirty years Gadbury collected daily observations of the weather, but failed to establish a pattern related to the motion of the planets. He tried to persuade the Royal Society to conduct statistical studies in other branches of astrology: 'why should not experiments in astrology be patronised by them?' (Capp 1979). The reaction of the members was sceptical. In fact, they refused to perform statistical experiments in astrology, thus demonstrating that in the 1670s those involved in experimental science were not likely to admit to an interest in astrology.

While Gadbury and others like him tried to establish an experimental basis for astrology as distinct from magic and the occult, they unintentionally hastened the end of its acceptability as a serious science at all. The word *science* in its modern sense is, of course, an anachronism: it was not used at the time except in the broad sense of *knowledge*. Nevertheless, the requirements for evidence and refutation that limit the modern definition of a 'science' were already taking shape. The more attention was focused on proving the claims of astrology, the more apparent became its failure to meet the requirements of the experimental philosophers.

At the same time, the invention of scientific instruments was also hastening the decline of astrology among the educated. The barometer could forecast the weather more effectively than the almanac. The telescope could show the imperfect, scarred surfaces of at least the Moon, and, later, of other planets. The appearance of new stars (*novae*) and comets, and the strange behaviour of variable stars, all indicated that the heavens could change. As well as demolishing finally the Aristotelian view of a changeless and perfect heaven, observing through a telescope became an activity of endless fascination with the constant possibility of new discoveries.

Some astrologers made serious efforts to incorporate the new heliocentric astronomy that was rapidly gaining ground in the seventeenth century into the main body of judicial astrology. The influence of Francis Bacon contributed not only an urge to collect statistical records of the relationships between predictions and events, but also to the making of more accurate tables of the motions of the heavenly bodies based on observation and the most up-to-date mathematics. In the mid-seventeenth century, among the best available tables in use in England were those made by Vincent Wing and Thomas Street. Both of these men were writers of popular almanacs, and neither had studied at a university, yet they were respected in academic circles as mathematicians and astronomers and their tables were used for astronomical observations completely unconnected with predictive astrology.

In spite of such links between astrology and astronomy, the gulf between the two was fully visible by the end of the century. Astrology continued to flourish but on its own separate course. The almanacs and horoscopes still provided the man in the street and the country lane with some hope of order in his environment and, particularly, a cheaper weather forecast than a barometer. Some academics also continued to dabble and to feel attracted to the power of the magus. An indication of this ambivalence can be seen in the great document of rationalism and faith in experimental science, the *Encyclopédie*, published in France in the mid-eighteenth century, in which astrology is listed along with physics in the table of natural sciences.

In England, John Flamsteed, the first Astronomer Royal, in spite of a certain lingering fascination with astrology which he was loth to acknowledge, devoted himself entirely after the founding of the Royal Observatory in 1675 to the study of the motions, nature and relations of the heavenly bodies. Although these could, in theory, include anything in the universe that could be seen, and did include such terrestrial phenomena as tides and effects in the Earth's atmosphere that would now be called meteorology, there were a few areas on which most interest centred during this period. To those who spent a large part of their time working in this field, as also to the educated population generally, there were questions that seemed at the time to be of overriding interest and importance.

The best known of these questions was how to find out the longitude. The essential difficulty for sailors was that, although they could find out where they were in latitude with the available instruments, no-one was able to keep the time at sea sufficiently accurately to make the calculations needed to establish longitude. The pendulum clocks available did not work well at sea. One possible solution was to use regularly recurring astronomical phenomena to indicate the time. The Moon was one possibility, but its motion turned out to be extremely complex. Another was the regular disappearance of Jupiter's satellites behind the planet itself. Making accurate tables of these motions became a major task for astronomers.

This sort of practical application must have seemed the main value of astronomy to outsiders. Understanding the motions of the Moon, for example, could lead to better tide tables and thus doubly contribute to more accurate navigation. Accurate astronomical observations could lead to a more satisfactory calendar. Interaction with the practical business of living went both ways. A request to reform the calendar for Pope Paul III had been the original stimulus leading Copernicus to examine Ptolemy's explanation of the movement of the solar system and the observations on which he based it. Different sciences and branches of sciences were then, as now, closely interactive.

Many who were interested in the problems of astronomy worked in the field sporadically, also spending time on other work. Christopher Wren, for instance, is best known as an architect, although he was an able astronomer in his youth. Robert Hooke is probably better known for Hooke's law in physics than for his astronomical work. Yet both of these men were sufficiently involved in astronomy to participate in the puzzles of the day.

Newton's *Principia*, published in 1687, answered a number of the main questions of astronomy at once. Nevertheless, sufficient challenges remained to stimulate much interest. The motions of the Moon and the particularly tricky planet, Mars, were still not fully understood. A large number of accurate observations would be needed before either could be adequately formulated.

The best observations ever made before the time of Newton were those of the Dane, Tycho Brahe. Brahe's work was extremely careful and highly accurate, surprisingly so since he worked without a telescope. He had reported the positions of planets and stars to an accuracy of 1 minute of arc. One of the tasks of later astronomers was to increase this accuracy. Flamsteed's telescopic observations achieved an improvement to the extent that his error was only about 10 seconds of arc.

With such accurate observing, Flamsteed felt ready to pose himself other questions and seek to answer them. In order to prove finally that the Earth does revolve round the Sun, it is desirable to show that there is an appreciable movement of the apparent position of the background of stars during the year when the Earth is at different points in her orbit. Devising an adequate method for observing this small movement and proving that it existed was one problem that he set himself. Another, with much more potential excitement for the general public, was to understand the nature and movement of comets. In 1680, a bright comet appeared, raising widespread interest and calling forth all sorts of speculations. In 1682, the comet which came to be known to succeeding generations as 'Halley's Comet' appeared. Halley was able to calculate and compare orbits of previous comets and correctly predict the return of his own in 76 years' time.

Even Halley, whose work in astronomy is probably better known to the layman than most, was involved in many other branches of natural philosophy. Flamsteed is unusual in the extent to which he concentrated on astronomy, but he was, after all, a professional for many more years than Halley. Newton, who became Lucasian Professor of Mathematics at Cambridge, spent years working in chemistry—and also a considerable part of his time not working in any science at all, as far as we know.

Among the amateurs who were involved to any great extent in

astronomy, the picture is similarly varied. Stephen Gray, whose work won the attention and praise of the Royal Society, began his scientific career by publishing work on a variety of subjects from microscopy to geology. He dabbled in almost all branches of science that were of interest to his contemporaries, and he died soon after winning the Royal Society's Copley Medal for work on static electricity. The only major field that he did not enter to any extent was medicine, and this can be seen as an accident of his circumstances. Others with the relevant background were able to combine medicine and astronomy in some ways, although James Pound and William Derham, both with medical degrees, concentrated their research interests in astronomy. Derham used a wide range of scientific material in his books *Physico-Theology* and *Astro-Theology*, in which he set out to prove the existence of God by the argument from design.

Astronomy claimed a high proportion of the time of scientists from all backgrounds. Among professionals, amateurs and paid workers, there were some who were involved in the most central and serious work of the time. Who these people were, how they came to work in astronomy and how they saw their work will form the subject of this book.

2

Astronomers

Even today, the label *astronomer* can be applied to a variety of different sorts of people. The professional researcher in front of his computer screen controlling a giant four- or five-metre telescope is an astronomer, and so is the amateur with a four- or five-inch telescope in his back garden. The careful amateur may still be the first to see an event like a nova, but he may no longer play much part in the routine collection of data. In the late seventeenth century, a gap between professional and amateur astronomers was just beginning to open. An amateur could still be involved in collecting data and in theorising over the main problems of the time, even though he might spend only a small part of his time on astronomy.

Across the strata of society, astronomy in the seventeenth century was accessible enough to allow not only aristocrats and landed gentry to participate, but also the middle classes, extending down to tradesmen and craftsmen. Essential requirements were as little as a grammar school education and some sort of instrument that need be nothing better than a very simple device for taking altitudes and an accurate clock for timing events. Although there is no way of assessing exactly how many were involved in astronomy in some capacity (see table 2.1), the numbers were so great that they became the objects of literary and artistic satire. New discoveries entered the popular consciousness to the extent that the great poets of the seventeenth century turned to astronomy, to new stars and comets, for images of change, doubt and uncertainty, as well as beauty. For many people outside science, there was much enjoyment to be gained from the insults of satirists like Jonathan Swift, Thomas Shadwell and Mrs Aphra Behn, who mocked the single-minded absorption of the astronomer with his eye always on the stars and no time for what seemed to be the more pressing social problems of the day.

7

Table 2.1 Some British astronomers active in the time of Newton.

Full-time professional astronomers
    1646–1719 John Flamsteed (Astronomer Royal 1675–1719)
    1656–1742 Edmond Halley (Astronomer Royal 1719–1742)

University men teaching astronomy or mathematics (includes Gresham College)

1630–1677	Isaac Barrow	1685–1731	Brook Taylor
1682–1716	Roger Cotes	1616–1703	John Wallis
1661–1708	David Gregory	1617–1689	Seth Ward
1638–1675	James Gregory	1667–1752	William Whiston
1635–1703	Robert Hooke	1614–1672	John Wilkins
1671–1721	John Keill	1632–1713	Christopher Wren
	1642–1727	ISAAC NEWTON	

Amateurs with income from land
      ?–?   John Godfrey
    1689–1728 Samuel Molyneux
    1628–1707 Richard Towneley

Amateurs with other jobs

	Church	Sea	Trade
1657–1735	William Derham	Henry Thomas	Stephen Gray
1667–1719	John Harris		Other
1669–1724	James Pound		
?–?	Stephen Thornton	1617–1679	Jonas Moore
?–?	Matthew Wright	1611–1685	John Pell
		?–?	Henry Stanyan

Paid workers in astronomy

Joseph Crosthwait	Thomas Smith
Cuthbert Denton	John Stafford
James Hodgson	Thomas Weston
Luke Leigh	John Witty
Abraham Ryley	Isaac Woolferman
Abraham Sharp	

Astrologers and almanac makers
    1622–1689 Thomas Street
    1617–1681 George Wharton

The picture of an astronomer that we might have after looking at the literary comment of the period would be of a member of the Royal Society with plenty of money and time on his hands to dally with useless speculation. Historians have looked at the whole membership of the Royal Society in order to form an idea of who the scientists were. They certainly confirm that the typical member in Newton's time was wealthy. Titled nobility formed a quarter of the membership. But historians who have studied the Royal Society as a whole have come away with conclusions that do not apply to the subgroup of astronomers. A close look at the men who were involved in astronomy at the time gives a very different picture.

Astronomers were not typical of the membership of the Royal Society in this period for several reasons. One is that the nature of the membership tended to change with the inclinations of the current President, and more importantly of the Secretary who was usually also the Editor of the Society's journal, the *Philosophical Transactions*. In his hands lay the selection of papers to be published in the only scientific journal in England. A study of the extent to which astronomy was published (see table 2.2) shows that at the beginning of the *Philosophical Transactions*, under Henry Oldenburg, astronomy was one of the best represented sciences, although many contributors, such as J D Cassini, were not English, and sent work in from abroad. Astronomy suffered a gradual decline in its representation, and under Sir Hans Sloane it dropped very far behind medicine and biology. The fact that Sloane was a doctor, physician to Prince George, may well have influenced his taste. Whatever the cause, the number of astronomical papers published by the Royal Society is misleading as an indicator of the level of activity in the country as a whole. Even though it was not fashionable at the Royal Society, many people were working in astronomy at all levels, even while Sloane was Secretary.

Table 2.2 Number of articles published in the *Philosophical Transactions* of the Royal Society (excluding book reviews).

	1665–6	1700	1731
Astronomy	35	3	2
Medicine	25	9	8
Physics	7	0	1
Mathematics	2	3	3
Art	1	1	0
Longitude, navigation	4	1	4
Natural history	32	11	4
Geology	28	0	1
Barometers	5	0	0
Geography, travel	11	6	0
Chemistry	—	2	0
Archaeology	—	5	0
Natural phenomena	—	—	5

The Royal Society's membership is equally misleading if used as an indicator of the social background of astronomers. From the founding of the Society until the death of Newton, the largest group of members, usually around 70%, came from the land-owning gentry (see table 2.3).

The titled gentry and nobility averaged about 25–30% and the small remainder came from the lower classes. The only exception to this pattern is the period between 1667 and 1682 when Robert Hooke was Secretary. Hooke, who had strong interests in astronomy, was a man who had to earn his living. His friends were craftsmen and working scientists, active in London. Under him, over 10% of new members came from below the level of gentlemen.

Table 2.3 Numbers of Fellows elected to the Royal Society.

	1660–1	1668–9
Aristocrats and courtiers	22	10
Gentlemen and country land-owners	8	6
Clergymen	5	1
Physicians	8	5
Tradesmen	3	5
Foreigners (mostly physicians)	0	12

Hooke and his friends are, in fact, much more typical of the astronomers of any period than they are of the Royal Society in general. In order to see who the astronomers were and where they came from, it has been necessary to look closely at the records of the period, mostly letters and memoirs which reveal details about individuals. From these sources some answers can be given to the questions: what sort of family produced an astronomer? Did the families help, and, if so, in what way?

The post of Astronomer Royal was the first full-time professional position in astronomy to be established in Britain. The post was, by its nature, a lonely one, and the character and upbringing of the first incumbent accentuated the difficulties. The position, originally called Astronomical Observator, and the Royal Observatory to go with it, were created by Charles II, following several years in which the plan had been in the air. John Flamsteed, who was eventually appointed first Astronomical Observator in 1675 (later to be known as Astronomer Royal), mentions a scheme for an observatory on a grand scale with 'an instrument as large as any the Arabs boast of'. Money was not immediately available and the Civil War had intervened to prevent any action being taken. The plan was shelved for a time, but, by 1674, the Royal Society was again considering the possibility of using their Chelsea College property to build an observatory.

At this stage, Sir Jonas Moore, Master of the Ordnance, became actively involved in the scheme. Moore was a mathematician who had developed an interest in astronomy, and who was able to use the resources of the Tower of London which went with his job. He had instruments and a library. He

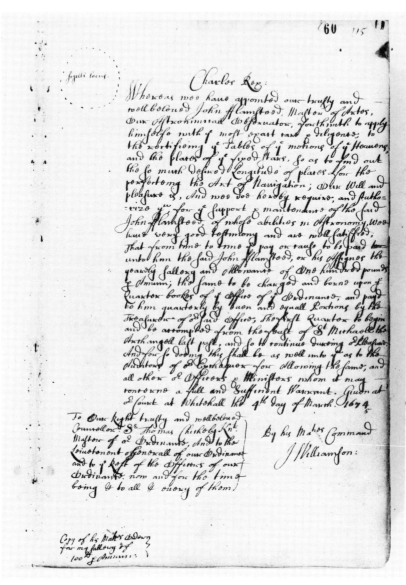

Figure 2.1 Flamsteed's copy of the letter by which King Charles appointed him Astronomical Observator with a salary of £100 per year. Flamsteed's note at the bottom suggests that he thought that he might need evidence of the promised salary. Royal Greenwich Observatory.

also had savings and a reasonable income and understood the need for accurate astronomical observations to solve the problem of longitude. When the King showed signs of renewed interest in the scheme for a royal observatory, Moore was prepared to offer to make a large contribution, or actually finance the building 'to the value of £150 or £200'. Moore was interested in being a patron of science, but had gained a reputation as a talker rather than an achiever. Robert Hooke, in his diary, wrote that he had heard a great deal of talk by Moore, and was exceedingly bored by it. Hooke did not expect anything to come of all the speculations.

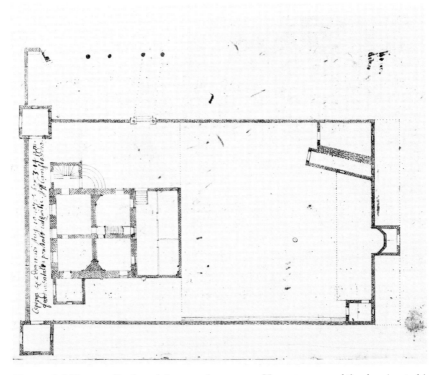

Figure 2.2 Flamsteed's plan of the new observatory. He sent a copy of the drawing to his friend Richard Towneley with details of the sizes of all the rooms and installations. Royal Greenwich Observatory.

Hooke was to be surprised. Moore *was* serious in his intentions and he knew the man for the job. In 1670, John Flamsteed, encouraged by his father, had made a visit to London from his home in Derbyshire. Flamsteed was a thin, nervous young man, but he impressed all those he met with his single-minded devotion to observational astronomy. He already had a sound knowledge of current astronomical theory and had begun to make observations of his own, yet he had never been to a university. He had come to London to talk to like-minded men and he managed to meet several members of the Royal Society, among them Sir Jonas Moore. The old man and the young immediately liked each other. Each respected the other's skills and abilities. From that moment, Flamsteed's devotion to astronomy never wavered.

Flamsteed's passage through youth is a case study of the openness of academic life at the time. Low social status and lack of formal education made life harder, sometimes made men bitter, but were no insuperable bar to anyone with something to contribute. Flamsteed was born on 19 August 1646, at Denby, in Derbyshire, whither his family had fled to avoid the plague. He was born into a family of tradesmen growing more prosperous in each generation, and his father owned lead mines. His ancestor was one William Flamsteed 'who came out of the north, bought the land at Hallam Mere, of one Robert Everet it being then rated 40s per year rent and died in 1514'. William Flamsteed had established the family in Derbyshire and was the furthest ancestor that John Flamsteed had traced when he came to write his memoirs.

Flamsteed's grandfather, William, of Little Hallam, who died in 1637, was described as a 'yeoman', an independent farmer under the rank of gentleman. His sons were increasingly successful. John, the oldest, 'improved the estate much'. His descendants took the title of 'gentlemen'. The second son, William, became town clerk of Nottingham. Stephen Flamsteed, the father of the astronomer, was the third son and he worked at the trade of maltster. He married Mary, who was the daughter of an ironmonger in Derby, John Spateman. Flamsteed said of his grandparents that they 'were of known integrity, honesty and fortune'. Flamsteed's immediate ancestors were tradesmen, but the family was sufficiently prosperous to allow him to attend the Free School in Derby and to continue there until he was 14. He states in his account of his life that he was prevented from going to university because of illness. At the age of fourteen, he was 'visited with a fit of sickness that was followed with a consumption and other distempers'. His father thought that the sickly John was not fit to go to university 'though I was designed'.

The result of this illness and enforced leisure at home was in Flamsteed's case the beginning of an interest in astronomy. But for much of his life, illness was a serious hindrance to him, causing him much suffering, which may go some way to explain the gloomy and humourless view that he tended to take of almost everything. The form of the illness that began when he was fourteen is described in detail: 'it pleased God to inflict a weakness in my knees and joints upon me . . . in the summer preceding, being bathing myself, together with some boys my companions (we might out of a general consent enter those baths which Lord Aston had erected on the side of the river) whence returning I found no hurt; but when I arose the next morning my body, thighs and legs were all so swelled that they would not admit me to get my usual clothes upon them.' This swelling was reduced by rubbing vinegar and clay on his body and legs. But, he points out, the symptoms were alleviated without the disease being cured. It went, instead, into his joints and caused a continual weakness, so that he was hardly able to stand, and had great difficulty in going to school.

Not being allowed to proceed to university was a great blow. At one point he notes with regret and a gentle rebuke his father's decision not to send him 'as, knowing the negligence of servants he might suppose that my presence at home might bridle, if not remove these disorders they were prone unto'. Mrs Flamsteed had died when John was only three years and two weeks old. She died leaving a month-old baby girl. Three years later Stephen Flamsteed took a second wife, Elizabeth Bates, but after giving birth to a baby girl, Katherine, she too died. The Flamsteed household thus had two young girls but no mistress to manage the servants.

Flamsteed understood that this might have been a reason for keeping him at home. His illness, he thought, was not sufficient reason. At university treatment would have been 'light and cheap'. His father, however, took notice of the violent headache that a day's reading brought on and assumed that a week's constant study would be intolerable. Flamsteed himself greatly regretted that he was not to go to university, being convinced that 'colds did oftener cause this disease than reading'.

In the country atmosphere in which the young Flamsteed grew up, the dividing lines between natural phenomena and magic were not clearly drawn. Belief in magic was widespread. His family took the practices of astrology and faith healing seriously, and, not surprisingly, these beliefs were part of Flamsteed's childhood. Some stayed with him, to contribute to his later habits of thought. Flamsteed's father owned a copy of Gadbury's almanac, published for astrological purposes, and this provided Flamsteed's first acquaintance with astronomical tables. More significantly, because of his illness Flamsteed became involved with Valentine Greatorex (or Greatrackes) who performed miraculous cures by touch. His fame spread throughout the country, sparked in 1665 by the apearance of a comet, which, Flamsteed notes, 'was much celebrated by the cures done in Ireland by Mr Valentine Greatrackes by the stroke of his hands without the application of any medicine' (Baily 1835). London coffee houses were full of debate, not so much over the efficacy of his cures, but over the mechanism by which they were achieved. The leading arguments suggested that some kind of 'effluvia' or 'friction' was responsible. The chemist Robert Boyle was involved in support of Greatrackes.

Stephen Flamsteed, however, was prepared to believe in Greatrackes' cures regardless of mechanism and, not wanting to miss any opportunity for helping his son, he arranged a visit to Ireland. His readiness to accept the possibility of a non-medical cure is probably related to his interest in astrology. Flamsteed himself, in assessing Greatrackes' work, was generous and not sceptical; he concluded, 'for my part I think his gift was of God'. Unfortunately, although Flamsteed says he witnessed some of the cures, he did not think he was qualified to describe them and so said nothing. For his own treatment, 'I was stroked by him all over my body but found as yet no amends in anything but what I had before' (Baily 1835).

The journeys were uneventful, except for some problems caused by fear of the plague, which was very bad in London in 1665. The passengers arriving in Ireland were closely questioned about their place of origin and, on the return, Flamsteed, having passed through Liverpool where the sickness also raged, was refused lodging at an inn, even though the innkeeper was a friend of his father. In spite of all, he returned home—safe but uncured.

In the winter of 1665/6 Greatrackes came to England and Flamsteed went to see him once more at Worcester, whether of his own accord or at his father's wish he doesn't say. The result, however, was the same. He was 'stroked by him' but with no better effect than formerly, though 'several then were cured'.

In 1675, Flamsteed became Astronomer Royal and embarked on several programmes of astronomical research. In 1688, he wrote a small note in the margin of one of his observing notebooks with measurements of lunar distances: 'On March 8th my father Stephen Flamsteed of Derby died aged 70.' By his death, Stephen Flamsteed made his most obvious contribution to his son's career. The death of Stephen left John with a considerably higher income than before. Managing the land in Derbyshire with the mining rights became one of Flamsteed's major worries. He took over his father's papers and had to learn the technicalities of the price of ore, the number of loads mined each year etc. Occasionally among his notes are records of court cases and various kinds of disputes and arguments.

When Flamsteed took over from his father, the mining concerns had expanded to a point where they needed constant supervision. Flamsteed already had his astronomy and the post of rector of Burstow. His heart was entirely in the astronomy and the family mining interests came a good way down on the list of ways he would choose to spend his time, but he was always inclined to worry. He wrote anxious letters to his friend Luke Leigh, who lived on the spot. In 1698, Leigh wrote to tell Flamsteed that a plot was being hatched to take control over a certain grove that belonged to the Flamsteed estate. Flamsteed replied that his father had owned several groves in that area and hoped for advantage from them when the drainage sough (or channel) reached them. Unfortunately, he had not had time to look at his father's papers in detail, and he asked Leigh to find out exactly where the grove lay and how much money would be required to gain a clear right to mine the area. It was, he realised, a great disadvantage to be an absentee landlord who could be so easily cheated. He hoped, however, that Leigh's honesty would help him there and he held out the hope of reward: ' 'tis probable it may turn to your advantage'.

Figure 2.3 This portrait of Flamsteed was painted by Thomas Gibson in 1712. Royal Greenwich Observatory.

Flamsteed's father's death gave him much anxiety, but the first effect was immediate cash so that he could pay for a new astronomical instrument known as a mural quadrant or mural arc, for taking the altitudes of stars. It was an instrument for which he had hoped and planned for years with many disappointments. More fundamental than this, though, his father's interest in books and education, even though the books were astrological and the education was incomplete, had allowed Flamsteed access to astronomy in the first place.

Whether as 'Astronomical Observator', or as 'Mathematicus Regius', as he also called himself, Flamsteed intended to devote the whole of his professional life to astronomy. There were other outstanding contemporaries spending part of their professional lives on astronomy and interesting comparisons in family background can be drawn.

Edmond Halley became Astronomer Royal when Flamsteed died in 1719. His background had much in common with Flamsteed's, but was more helpful to a future astronomer in nearly every way. Like Flamsteed, Halley was the son of a prosperous tradesman, Edmond Halley of Haygerston in East London, now known as Shoreditch. The Halleys owned a soap boiling business in Winchester Street. The grandfather, Humphrey Halley, is described as both a vintner and a haberdasher; thus, alcohol played a part in founding the fortunes of both the Flamsteed and Halley families.

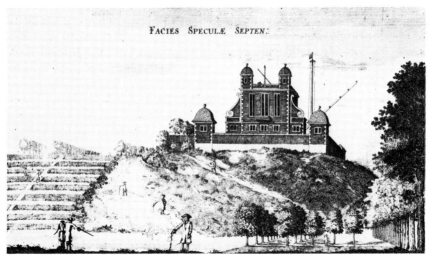

FACIES SPECULÆ SEPTEN.

Figure 2.4 One of Francis Place's engravings showing the Royal Observatory as it was when Flamsteed inhabited it. This engraving shows the north face. Royal Greenwich Observatory.

The Halleys were considerably richer than the Flamsteeds when the two boys were born. Halley's father had rents of over £1000 from London properties as well as the income from his soap business. Yet his establishment did not provide a tutor for his son. John Aubrey (1898) says that Halley was taught 'reading and arithmetique' when he was nine by one of his father's apprentices.

For the first part of his life, however, Halley lived comfortably, thanks to his father's prosperity. He was sent to St Paul's School, and later to

Oxford, where he was able to spend time on astronomy. It was already his great passion. Although his father does not seem to have played much part in creating Halley's interest in astronomy, he seems to have tolerated it and even given considerable financial support before disaster overwhelmed him.

Halley was so involved with astronomy that, as Aubrey (1898) says, he left Oxford 'before he was of standing to take his first degree and finding his father willing to gratify his curiosity and furnish his expenses necessary for his Instruments and Voyage, he, after thinking of several places at last pitch'd upon the island of St Helena, a settlement of the East India Company'. Halley's father sympathised with his son's adventurous, if unorthodox, plans to the extent of allowing him £300 per year. Aubrey says 'he got leave and a *viaticum* [travelling allowance] of his father to go to the Island of *Sancta Helena* purely upon account of advancement of Astronomy'. Aubrey obviously thought that Halley's father was unusually generous to his son's probably non-profit-making ambitions. Status and public recognition were also of value to successful businessmen and the fame brought by the voyage was welcome. This may have been an incentive to Mr Halley.

Although Halley's life had begun with every encouragement for him, yet while still a young man he suffered several tragedies. Many of the astronomers of this period lost one parent in early youth. Halley lost his mother, Anne, who died in 1672. Materially, the family had suffered a heavy loss back in the Great Fire of 1666. The property in the City owned by the Halleys was badly damaged. To make matters worse, Halley's father, like Stephen Flamsteed, remarried. His second wife appears to have been a difficult woman and one who had no great liking for her stepson. The marriage was never very happy.

The Halleys' financial affairs continued to go from bad to worse. According to a broadsheet now in the Guildhall Library, on Wednesday 5 March 1684, Halley's father disappeared. That morning he had complained that his shoes were hurting him and, to ease his feet, his nephew cut out the linings. This seemed to help and he told the family that he would go out walking, but be back in the evening. He set out and never returned. Mrs Halley, having lost the bread winner, offered a reward of £100 for information of his whereabouts. She did not have to wait very long. On 14 April a gruesome discovery was made. By the riverside at Strood in Kent a boy, John Byers, discovered the body of a man, naked except for his shoes and socks. He told a passing gentleman, a Mr Adams, who, remembering the mysterious disappearance, sent to Mrs Halley. The face of the body was quite disfigured but, by his cut-up shoes, it was identified as that of Halley's father. A coroner's jury 'brought him in murther'd' but Macpike (1937) suggests that it was more likely to have been suicide, because of the problems that he had been facing. His end was made more sordid by a squabble between Adams and Byers over the reward. The gentleman had at first said that the boy should have all the money but later changed his mind. He brought a suit for the whole £100. The judge awarded him £20 and the remaining £80 went to the boy, John Byers, who had actually found the body.

Halley had been in Italy on a grand tour (and therefore was apparently still able to spend considerable amounts of money) when he was called home in 1681 because of his father's financial problems. From then on he received very little money from his family. In fact, the state of the family relationship can be guessed as he instituted a law suit against his stepmother to recover what he could from his father's estate. Halley, from then on, was compelled to earn his living. He was one of the fortunate few able to live from science. He worked for the Royal Society in a semi-scientific post, then in a university, and ultimately became a professional astronomer: Astronomer Royal, when Flamsteed died.

Figure 2.5 Edmond Halley. Engraving. Royal Greenwich Observatory.

A brilliant inventor, a creative scientist and astronomer but an infuriating personality, Robert Hooke, had much less financial support from his family than either Flamsteed or Halley, and he was worse off than both in that he lost both his parents while he was a boy. He was the son of the Reverend John Hooke, curate at Freshwater in the Isle of Wight. Like Flamsteed, Hooke suffered from bad headaches even as a child and so was never sent to school. He spent his time at home creatively, having already developed an interest in mechanics. Aubrey tells us that by himself, as a child, Hooke invented 'a wooden clock that would go' and a model of a ship that could fire off small guns. Perhaps his father taught him some mathematics or mechanics, or perhaps he learnt from craftsmen in the village. In October 1648 when Hooke was a boy of thirteen, his father died, 'by suspending himself', according to Aubrey. He left to his son £100. With this, Hooke set off to London on his own, physically weak but already showing remarkable strength of character. He enrolled himself at Westminster School under Dr Busby and from there went to Christ Church, Oxford, as a chorister. For the rest of his life he had his living to earn and his diaries, full of notes on expenditure and earnings, show how conscious he always was of the passage of money through his hands.

Hooke at least kept himself fed and never suffered great hardship, as far as we know. Others were worse off. One of the most extreme cases was the physician and economist William Petty who was left with a legacy of only £10 when his father died—and that was never paid. He went to sea, but was put ashore by the crew of his ship on the coast of France because he had a broken leg. He turned this hardship to his advantage by enrolling in the Jesuit School at Caen. Aubrey gossips 'that he lived a weeke on two pennieworth (or 3 I have forgott which but I think the former) of wallnutts'.

A large group of astronomers worked for all or part of their careers in universities. This group will overlap with other groups as many men involved in astronomy spent at least part of their time working at one of the universities, or at Gresham College in London. This in itself is interesting, as it has been customary for historians to question the importance of the universities to science during the seventeenth century, and to argue that important work was done mainly outside universities. Not only is the proportion of astronomers who worked in the universities fairly high, but these men form a group with other characteristics in common.

A large number came from families connected with the Church or with land-owning. Roger Cotes, James Gregory, Robert Hooke, William Whiston, Christopher Wren and John Wallis all had fathers in the church. Hunter (1981), in a study of the British-born scientists of the late seventeenth century who were included in the *Dictionary of Scientific Biography*, found that 23% were sons of Anglican clergy. This is a high proportion, as sons of clergy form only 1% of the general population. He suggests that the marriage of the clergy and consequent production of sons was one of the main influences on the growth of interest in science. This argument is only partially born out by a study of astronomers as a separate group since many of their fathers died while they were young. All that they received was a genetic effect, although, in some cases, they inherited a library. Several were influenced instead by uncles who were involved in the Church. One of those whose fathers played a minimal direct role is Newton himself.

Newton came from a family of yeomen farmers who had become more and more prosperous in land-owning activities in the area surrounding

Figure 2.6 Portrait of Isaac Newton by Sir Godfrey Kneller. Royal Greenwich Observatory.

Grantham. Newton's ancestors had large families of sons, all of whom shared in the divisions of land and most of whom had farmed successfully. Newton's father, Isaac, inherited the Manor of Woolsthorpe which his own father, Robert, had bought. The Newtons were thus doing very well by yeoman standards. However, Westfall (1981) points out that not one member of the Newton family had been literate. All legal documents were drawn up by lawyers and clerks and signed only by a mark. In such a family the sons might have been sent to school, but there was certainly no background of learning. Isaac was fortunate, however, in that his father married Hannah Ayscough (or Askew); her family was prosperous and had also begun to move towards a middle-class interest in advancement in education. Hannah's brother William Ayscough was admitted an MA at Cambridge in 1637. He became rector of a nearby parish in the year in which Isaac senior married Hannah.

Isaac Newton senior died after six months of marriage. The only surviving description of him says that he was 'wild, extravagant and weake'. His influence, whatever it might have been, was removed before his son was born, and the result was that the uncle took it for granted that the uncommonly bright boy should go first to school and then to Cambridge.

The illiteracy of the Newtons, including the cousins, suggests that Isaac Newton might have been held back by his father if he had lived. We can only guess how an individual life would have evolved in different circumstances, but we recall that others with minds of potential power, notably Robert Hooke and William Petty, managed to find an education for themselves without father or uncle.

Some astronomers teaching and working in the universities in the seventeenth century came from the families of craftsmen. John Wilkins, at one time Warden of Wadham College, Oxford, and the author of *Mathematical Magic*, an attempt to make geometry and the useful aspects of astronomy available to a wide audience, was the son of a goldsmith, Walter Wilkins. His parents, however, combined the two kinds of background from which astronomers frequently came: the crafts and the churches. As his mother, Jane Dod, was the daughter of a Puritan divine, it is not surprising that Wilkins was enrolled at Magdalene Hall with its Puritan association. He was able later in life to apply the craftsmanship learnt from his father in such projects as working with Christopher Wren on building an 80-foot telescope designed to allow observation of the whole face of the Moon at one time. This telescope was never successfully completed, but many smaller ones were put together and used for practical observation.

Also working in the first half of the seventeenth century was the mathematician William Oughtred, the son of a butler at Eton College. The older Oughtred was, however, a butler who cared about education, and he provides one of the clearest cases where a father taught his son and stimulated an interest in learning. Oughtred's father taught him to write and to do common arithmetic. Eventually he secured a scholarship for his son at Eton. Oughtred himself was always highly regarded as a teacher and 'he taught all free'. Unfortunately 'none of his sonnes he could make scholars.' Perhaps they suffered from the rigorous regime of Oughtred's wife who was 'a penurious woman' and would not allow him 'to burn candle after supper'. His oldest son reversed the trend from craft to science by becoming a craftsman, a watchmaker.

The Gregorys are an interesting family. James Gregory, who invented the reflecting telescope that bears his name, was the son of the Reverend James Gregory, Minister of Drumoak in Scotland. Through his mother he was related to the Andersons, a family that produced mathematicians and mechanical engineers. A relative of his mother was the famous 'do it a' Anderson'.

The principles of 'do it a' ' were common to various branches of the family. David Gregory, James's nephew, absorbed it to the extent that he became Professor of Mathematics at St Andrews when his uncle left, and eventually Savilian Professor of Astronomy at Oxford. David had in fact read his uncle's papers and from them derived his interest in astronomy and particularly in the working of the telescope.

A few serious astronomers had fathers who could be considered aristocratic or belonging to the landed gentry: Laurence Rooke, Brook Taylor, Richard Towneley, George Wharton, Samuel Molyneux, John Keill, James Pound. David Gregory's father had worked for a merchant but inherited the family estate on his brother's death. The paths of all these men were eased by inheriting comfortable amounts of money. In some cases, too, the father or a close male relative gave direct teaching or encouragement in scientific or mechanical subjects or mathematics.

Of those who made important contributions to astronomy during the late seventeenth and early eighteenth centuries, probably the nearest to genuine aristocracy was Richard Towneley, but as he was a Catholic he suffered severe limitations on his freedom. The Towneleys had held estates in Lancashire since the thirteenth century. They had continued to prosper and acquire more land since then, but had run into trouble in the sixteenth century. In the reign of Elizabeth, in 1581, John Towneley was imprisoned for refusing to give up Catholicism. Undeterred, he made use of an influential relative who was at the time Dean of St Paul's. He was released from prison, and he and his family remained Catholic. They were notorious for harbouring priests: a folding altar was kept at the land agent's house nearby.

In the seventeenth century the Towneleys were Royalists. Many

Catholics from the northwest joined the King's army, and Charles Towneley petitioned to be allowed to keep armed bands for self defence; but his petition was refused. The family became known for courage and conviction. To such a background of independence and determination the Towneleys in the mid-seventeenth century added an interest in science. Christopher Towneley was well known as an antiquarian with a desire to collect almost everything, including the scientific papers of some of his colleagues who had died in the Civil War. He was responsible for the preservation of the letters of William Gascoigne, inventor of the filar micrometer, and also those of William Crabtree and Jeremiah Horrocks.

Round him at Carr Hall in the Forest of Pendle, Christopher Towneley assembled a small group of northerners with interests in science and astronomy. His nephew, Richard, inherited the main Towneley estate and grew up in the disturbing days of the Civil War. The Towneley family suffered a good deal. Aged thirteen or fourteen, Richard lived through the sequestration by Parliament of all the family estates except Hupton Manor. During these years, Richard was taught at home by a private tutor, and must therefore have felt the full brunt of the family anxiety and insecurity. In later years he showed no desire at all to be involved in politics or religious controversy, and, instead, devoted his attention to science.

The Towneleys, however, did better than many other Catholic families. In 1653 their lands were restored to them. Richard was able to make use of a servant in his astronomical observations, a luxury for which some of his colleagues would have been very grateful. His family background ensured not only wealth but also the opportunity to take an interest in the intellectual ideas of the time. He probably visited the continent as a youth. His younger brothers Charles and John were certainly sent to the Jesuit College at Douai in Belgium in 1649, and must have brought back news of ideas from the Catholic universities where the influence of Descartes was most strong. For a Catholic who lived in England, however, the times were dangerous. In 1678, two Jesuit priests called Pickering and Ireland were condemned to death for plotting the murder of James II. Flamsteed writing to Towneley in 1671 sympathises with him: 'I am sorry the present state of affairs will not permit you a journey to London, nor make any distinction betwixt the jesuitical Cabalists and others. These every day endeavour to raise new troubles and expiate their late crimes with greater, which has so exasperated the Parliament that all of the Romish Persuasion will be more severely confined than by any former Acts.'

Flamsteed adds that 'enforced leisure at home may at least lead Towneley to be more with truth, the heaven and the stars'. If Towneley could not come to the Observatory, the Observatory 'shall visit you the oftener'.

While Richard appears to have stayed at home, taking Flamsteed's advice, his brother Charles was still travelling, and in 1678 Flamsteed asked whether Charles had yet returned from Paris because he wanted a book forwarded.

Towneley, therefore, had a source of books, connection with London as well as his local circle of friends, and probably did not suffer greatly from isolation in spite of his Catholicism; perhaps in some practical ways he was helped by it. He was able to make observations and later to form the nucleus of one of the earliest groups of collaborative scientists in the seventeenth century. The groups of scientists surrounding the Towneleys were of great importance, both because they were collaborating and because they perpetuated a belief in observations as the heart of astronomy.

Christopher Wren was another astronomer who suffered for his family's religion. He was a university man, son of a cleric, the Reverend Christopher Wren, later Dean of Windsor and chaplain to Charles I. His uncle was Matthew Wren, Bishop of Ely, but, in the Civil War, the orthodox Anglican family was divided. Matthew was arrested and imprisoned in the Tower. The young Christopher took refuge with William Holder, an Oxfordshire mathematician who later became his brother-in-law. This turned out to be an important advantage to a boy with Wren's potential. Holder taught him mathematics and astronomy, establishing an interest which increased later at Westminster School and Oxford, where he was at Wadham College under John Wilkins. Yet again the family background gave rise to the crucial relationship, but it was not the father himself who gave the boy his interest or his knowledge.

In his study of the social background of seventeenth-century scientists Hunter (1981) emphasises the 'relative fewness of merchants and artisans' sons', and says that, even if science had become middle class, it was not bourgeois. Scientists with an interest in astronomy certainly often descended from clerical fathers, but merchants and artisans were more common as fathers if the group is extended to include amateurs and paid assistants. Among them there are, as might be expected, a great number of artisans' and merchants' sons.

The assistants of John Flamsteed, working in all sorts of capacities as computers, calculators, as observing assistants, or as independent observers supplying him with extra material for his tables, came predominantly from the families of tradesmen. In his years as Astronomer Royal Flamsteed employed at one time or another at least eleven paid servants who worked at the Observatory. For most of them we have to guess the family background. Cuthbert Denton was a 'surley labourer' who never developed an interest in astronomy but seems always to have found his work a burden. We hear from Flamsteed that Hooke's badly made wall quadrant 'had like to have deprived Cuthbert of his fingers'. On another occasion he committed the worst possible sin for an astronomer's assistant. He ran off to London and did not come back for two nights, both of which, Flamsteed bitterly points out, were clear. Cuthbert does not seem to have had much education. He never seems to have tried to learn mathematics, and his work never progressed beyond labouring. Others came much better prepared, or at least developed an interest after they arrived. Flamsteed was always willing to teach and help assistants to develop their own strong interest in what they were doing.

Luke Leigh, a 'poor kinsman' of Edmond Halley, was 'skilled in mathematics'. James Hodgson, who later married Flamsteed's niece, was 'a sober young man about 22 years of age. A very good geometrician and Algebraist. He understands the series and Fluxions & . . . the latin tongue'. Joseph Crosthwait was a 'Cumberland youth', possibly the Joseph Crosthwait who was baptised at Causey Head on 26 November 1691. He must have felt himself very far from home while, at the age of thirteen, he laboured over his calculations at Greenwich.

Flamsteed had close and affectionate relationships with several of his assistants. His relationship with Abraham Sharp is particularly clear to us because when Sharp returned to his home in Yorkshire the two men corresponded for the rest of Flamsteed's life. Sharp was born at Little Horton, near Bradford, in 1653. His father was a clothier who also owned a little land. There seems no evidence in his childhood to suggest how he might have developed an interest in mathematics or astronomy at home. He was lucky, though, in that his father sent him to Bradford Grammar School after he had exhausted the potential of the village school at Little Horton. The father's intention was no doubt that Abraham should take over the family business and, in 1669, when he was sixteen, Sharp was apprenticed to a mercer, or dealer in textiles, in York. A premium of £20 was paid for the apprenticeship, but in spite of all parental plans and pressures he was so unhappy that he left to become a teacher.

While he earned his living in this way, Sharp was able to spend most of his time on the study of mathematics, which was what he really enjoyed. Through his studies he came to hear of men who were making a name for themselves in London, among them, of course, Flamsteed. He is said then to have found a job as accountant to the merchant in whose house Flamsteed was living at the time, and so made his entry into the world in which, clearly, his interests already lay. Whether this is true or not, we do know that Flamsteed's assistant, John Stafford, died in 1688 commemorated by a little note in the margin of the observing notes: 'On May 28th at 7.30 a.m. Stafford died, my servant in observations for 3½ years.' In his place, Sharp, who had already come to Greenwich in some unofficial capacity, was appointed officially as Flamsteed's assistant.

Flamsteed found that his major task of producing an accurate catalogue of positions of fixed stars involved an immense amount of observation, calculation and, above all, painstaking accuracy in checking and rechecking every result. In addition to this, his main life's work, he was also engaged at various times in work on stellar parallax, the motion of Mars, sunspots, a new lunar theory with tables, more accurate tables of all the planetary orbits, and, especially, the orbits of Jupiter's satellites. For all this work, Flamsteed was obliged to look for help beyond the amount that could be done by the one or, at most, two paid assistants that he could afford to keep.

He solved the problem partly by paying some calculators to work at home checking against results obtained at Greenwich. Luke Leigh and Abraham Sharp were both employed in this way at times and were very grateful for even the small sums which they occasionally received. In addition to these, however, Flamsteed managed to build up a network of interested amateurs who were willing to send him their observations, free of charge, for the honour and glory of being involved in science.

The backgrounds of these men might be expected to be largely aristocratic. We might expect country parsons also to form a large proportion of the sort of men who would wish to practise astronomy and would have the knowledge and the equipment to do so, not to mention the sort of working hours that would allow devotion to such a hobby. In fact the majority of Flamsteed's correspondents consisted of clergymen. The Reverend Matthew Wright of Crewe in Cheshire signed himself 'an unworthy priest of ye Church of England'. He introduced himself in a similarly humble tone: 'Worthy Sir, Pray pardon this trouble given you by an obscure person who never had ye honour to be personally known to you but who would count it ye highest pitch of his Ambition to be reckoned in ye number of your Mathematical Correspondents' (22 December 1716).

Other clergymen included the Reverend Stephen Thornton, who kept a school at Ludsdown in Kent, and the much better known Reverend William Derham, who began his career with the help of a patroness, Lady Grey, but ended as chaplain to the Prince of Wales and, later, as a canon of Windsor Chapel with no particular need to be humble to John Flamsteed.

Another group of correspondents included sea captains and voyagers. James Pound combined both vocations. He was sent to Oxford in 1687 at the age of eighteen. He graduated in medicine, took orders and set sail for the East India Company as a chaplain. He had an eventful time in southeast Asia, but managed to write some letters to Flamsteed. His letters show the frustration he must have felt waiting for three years for a quadrant to be sent out to him and waiting in vain for Flamsteed's tables of the movements of Jupiter's satellites so that he could contribute to the observing programme. In fact, the tables never came and, anyway, because of a native uprising, Pound lost all the observations that he had managed to make.

Pound's first successful practice of astronomy began after his return to

England when Halley introduced him to the Royal Society as already well skilled in astronomical observation. The voyages may have helped him to acquire the practical skills of astronomy, but the interest was there before.

Flamsteed taught many pupils who later went to sea. At various times of his life he therefore received letters from all over the world, both from men whom he already knew, and from some who introduced themselves and proceeded to correspond partly for their own benefit. In spite of the opportunities for astronomy at sea and abroad, Flamsteed received few valuable observations from his correspondents. John Collins, for example, spent seven years learning mathematics and navigation while serving the Venetians in their war against the Turks. Yet Collins never became an active astronomer, preferring to facilitate correspondence between others. Thomas Brattle wrote from New England and might have provided useful data. He suffered, like many others, from lack of specialist knowledge and accurate instruments.

On the whole, men who stayed working steadily at home produced the most useful results. Of these, one of the most humble in his own estimation, connected neither with the church nor with the land, but the owner of a dyeing business, was Stephen Gray. He was, nevertheless, able to produce useful data. Stephen Gray wrote a number of letters to Flamsteed in which he offered, or apologised for not offering, observations on eclipses, the moons of Jupiter, sunspots and, once or twice, data on other planets. The fascinating question is how the interest in careful, routine astronomical observations developed in a tradesman living in Canterbury.

Gray's case can be taken as an illustration of the way in which scientific interest could allow a degree of contact with the highest social circles. Gray's life, as far as we can piece it together from the fragmentary evidence that survives outside his own letters, also shows a high degree of interaction among social classes at Canterbury.

Like so many other astronomers, both amateur and professional, Gray grew up in a family which does not at first sight seem a likely place to nourish a scientist. The Gray family had lived in Canterbury for generations. The first direct ancestor that we have been able to trace is Henry Graye, a blacksmith at the end of the sixteenth century. In 1599 he bought, for 30 shillings, the freedom of the city to trade as a blacksmith. From this time on, the family took on various trades and seems to have been increasingly prosperous. This is certainly a feature shared by many astronomers of this group. The families of Flamsteed, Halley and Newton all show similar upward movements in wealth and possessions, although these changes were not always accompanied by any apparent interest in education or in moving into the professions.

In the Gray family, the trade of blacksmith lasted for another generation through Richard Graye, who was baptised in 1609. His son, Matthias Gray, was the first member of the family recorded as a dyer. The freedom of the city acquired by Henry Graye was handed down from father to son once acquired and continued for the dyers as it had for the blacksmiths.

Matthias Gray married Anne Tilman in Canterbury Cathedral and settled in a house in Best Lane; the two had seven children. Two of the girls, Mary and Elizabeth, died as children. The oldest son, Thomas, entered the dyeing business with his father. The second son, Matthias, became a grocer, and the third, John, became a carpenter. The fourth and youngest son was Stephen Gray. He was baptised at All Saints' Church, Best Lane on 26 December 1666. His actual date of birth is unknown but it was probably just a short time before. The infant mortality rate was such that baptism was not left long after the birth unless there were special problems. All Saints' Church in Best Lane has been pulled down. Over the site now stand the gas company's showrooms, but, in a little grassy space to the side, some of the old tombstones still stand. Unfortunately, most are too weathered for the inscriptions to be legible. The name 'Gray' does not appear on the legible stones.

The Gray dyeing business moved from Best Lane at some stage when Stephen was a boy. A better position was found near the river in Stour Street, and the family moved while Stephen was still young. He stayed to work when his brothers moved out. In fact, dyeing was carried on in Stour Street until the 1960s, in the same shop.

Only two boys in the family, Thomas, the oldest, and Stephen, the youngest, entered the dyeing business. The work was physically very demanding and seems to have exhausted Stephen, especially as he grew older. Thomas died in 1695, leaving Stephen with much of the heavy work to do himself. By the time he was forty he was suffering from various pains, especially in his back. 'I have been much afflicted with the Dolor Ischiadis for near a quarter of a year but thanks be to God am now Almost free from it feeling noe pain except I attempt to labour hard' (1 May 1706). Five years later, his back was still weaker and he was suffering from 'a strain I received in my Back some years agoe which brought on me the Dolor Coxendicis' (31 July 1711). He speaks at this time of the 'difficulty and pain' caused by the work as being 'more than in former years'.

In spite of all his effort, Gray never seems to have made much money from dyeing. His brother, Matthias, the second son who became a grocer, was more prosperous and his civic activities show that he was working into a position of authority in the town. Gray pointed out, with obvious pride, that the Matthias Gray who became a councillor in 1688, an alderman in 1691 and finally Mayor of Canterbury in 1692 was actually his older brother.

Whether Gray owed any of his interest in astronomy to his older brother is not clear but he may well have owed him some of his introductions to members of the scientific community. Matthias was a personal friend of the Astronomer Royal. In 1708 Stephen wrote to Flamsteed: 'My brother comeing to London designed to make you a visit but his bisiness falling out crossly was disappointed' (18 September 1708).

Because Stephen's correspondence with Flamsteed had begun in 1699, there is no clear evidence that Matthias was the initiator. We do know that it was through Matthias that Stephen became involved in one of the major scientific disputes of the day. Matthias had married the daughter of John Somers and had thus come to own some land out at Chartham, near Canterbury. In 1701, Gray wrote to Hans Sloane at the Royal Society about some bones that had been found in a well on the Chartham estate. There was a jawbone and some teeth, and in the light of the controversy among Burnett, Woodward, Halley and others over the historical reality of the Flood and over the origin and history of the Earth, the position in which such bones were found was of great interest. Gray presented his speculations on whether the bones could have been deposited before or during the biblical flood. He also considered the likelihood that the bones belonged to an elephant or mammoth.

In reply, apparently to a query from Sloane, Gray wrote proudly of the discovery and of the old gentleman, Mr Somers, being lowered down the well in a basket in the interests of science. Gray was certainly proud of his family's new social position: 'What bones and drawings of them that were dugg up at Chartham are in the hands of Mr. Alderman Gray whom you suppose I must needs know which was a little pleasing to me he being my own Elder Brother. His present wife was the wife of Mr. John Sommers' son for whom he built the house where these bones were dug up. She says she was there when her father Sommers was let down into the well by a Basket and saw the Bones soe soon as brought up and that not longe after they were all except one tooth put up into an oval wooden Box and sent to Oxford haveing first caused Limmer to draw some of the Most Remarkable ones.' Matthias Gray had become a land-owner, and a further indication of his prosperity is that he sent his son, John, to Cambridge, to study medicine. Education was valued and seen as a means of advancement for the next generation, although neither Stephen nor Matthias was able to go to a university and follow a profession.

Meanwhile, Canterbury provided Stephen Gray with other means of advancing his scientific career. Another young man with a great desire to work in the widening fields of science grew up in Canterbury at about the same time as Gray. His name was Henry Hunt. Unlike Gray, he made no effort to work in Canterbury, but went to London, where he began work as the assistant of Robert Hooke. Hooke records in his diary for 'Jan 1672/3 Harry first out of the country. Agreed with Harry for £10 per annum.' From then on Hooke refers to Harry frequently, but often in a less than complimentary tone. No doubt Hooke was a difficult person to work for.

His health was poor. He suffered from sinus headaches and stomach troubles and dosed himself with every medicine that he heard mentioned. The result was considerable physical pain and a short temper. As often happens, Hooke found everyone else difficult to endure. He wrote on 24 March 1673: 'Harry surly and proceeded to put on hat—he shall march.' Harry may have walked out from time to time but he returned and, in fact, worked as assistant to Hooke for some years.

His work must have involved learning a good deal in many fields, a kind of scientific apprenticeship. Hooke wrote: 'with Harry I wrought on the specular metal for telescope and polish't it pretty trew' and 'Harry set up a bench and melt stones for furnace.' 'Harry cleaned lathe.'

In 1676, Hunt was appointed 'Operator' to the Royal Society at a salary of £20 per annum. He worked for the rest of his life for the Royal Society in various capacities, and, although his relations with Hooke varied, they seem to have been fond of each other.

Since most of the scientific practitioners of the 1670s and 80s knew Hooke and were likely to be involved with the Royal Society in some capacity, Harry Hunt was increasingly in a position to understand what was going on. He never married and there is no evidence to tell whether he made many visits to Canterbury once he had gone to London. Whether he did or not, he was certainly known to Stephen Gray, and it was through him that Gray first made contact with the Royal Society. We do not know how the correspondence began, but letters were already going back and forth when Stephen Gray's first extant letter was written in February 1696. The tone is formal and suggests an acquaintance rather than a friendship: 'Sir, I heer return you the account of these Optic Physiological Experiments and Observations which the time I could spare from my more necessetous Avocations for a livelyhood have this winter Permitted me to make' (3 February 1696).

It is clear that Gray felt that communicating with Hunt was more appropriate than writing to Secretary Sloane direct. He asks Hunt to 'communicate to the learned Secretary from his and your Humble Servant'. Hunt evidently did this, and, by March 1699, Gray had become sufficiently confident that his work would be welcome to write directly to Sloane.

The bulk of Gray's astronomical work, however, was communicated directly to the Astronomer Royal, John Flamsteed. The first extant letter from Gray to Flamsteed is dated 4 November 1699. This correspondence, too, was already established: 'Reverend Sir, In compliance to your desire I here send you the Construction of the suns eclipse in 1715 calculated from your Tables. It wants but a few seconds of being total at London as you see by the construction but with us at Canterbury it will be total' (4 November 1699).

We therefore do not know how Gray came to know Flamsteed. Hunt or

Sloane at the Royal Society may have put them in touch with each other.
Possibly Matthias Gray was the instigator of the correspondence.
Apparently, he already knew Hunt in 1696 when Gray's first known letter
was written: 'I . . . Have since my Brother was with you chanced to light
upon an other Experiment which I have incerted . . .' Later on we find that
Matthias knew Flamsteed and was in the habit of travelling to London
which, as far as we know, Gray did only rarely.

Flamsteed may have had family connections near Canterbury. A letter
from Gray in January 1716 shows that he was staying at Norton near
Faversham and working with the owner, Mr John Godfrey. In this letter he
mentions 'My honoured friend, your Cosen Godfrey who commands me to
present his Humble Service to you and Madam Flamsteed' which suggests
some close relationship between Godfrey and Flamsteed. Since, however,
Gray never mentions Godfrey before 1715, it seems unlikely that he knew
Godfrey first, and more probable that Flamsteed introduced them.

Since Flamsteed took private pupils in astronomy in an effort to
supplement his meagre salary, the possibility exists that Gray actually went
to Greenwich at some time to learn mathematics and astronomy (see
chapter 3). This seems unlikely because Flamsteed wrote a list of 'my
pupils as far as my memory will serve me'. Gray's name is not included in
the list, but Flamsteed was growing old by that time and might have
forgotten him.

A picture emerges of Gray working in his dyer's shop, waiting each day
for the time when he would stop his 'necessetous Avocations' and get on
with the work that really interested him. From his family and friends he
somehow picked up the necessary threads that would lead him to the Royal
Society, Flamsteed, and a close connection with the major investigations of
the day in physics and astronomy. How he and his contemporaries
acquired the knowledge, books and instruments that they needed will be
examined in the following chapters.

3

Preparation for astronomy

Astronomy can be practised at different levels. Assistants who carry out assigned tasks in calculation and instrumentation need a different sort of preparation from the scientists who originate research programmes. In the seventeenth and early eighteenth centuries, there were no clearly marked avenues to astronomy. The ways in which the theory and practice were learned show the development of new concerns in society and the presence of groups able to advance in new directions by adapting, in most cases, institutions that were already in existence.

To practise astronomy, more than interest is needed. A theoretical knowledge of mathematics and optics has become necessary. Moreover, for much of the seventeenth century, the tools of the trade, the instruments themselves, were made, lenses were ground, and adjustments were made by the astronomer who was to use them. Preparing the instruments thus required practical mechanical skill as well as optics and mathematics.

By the early eighteenth century, instruments of various capacities could be bought or made to order. Even so, anyone who wished to contribute significantly to astronomy needed mechanical knowledge to set up and adjust instruments and, of course, basic knowledge of the work of classical astronomers, as well as of all the developments of the seventeenth century.

The traditional picture presented by historians is that schools and universities were of little use to scientists. A close look at the life stories of astronomers shows that some of them did acquire a great deal of useful knowledge at school and even more at university. There is, however, no denying that a few scholars outside the universities who were also good

33

teachers, like William Oughtred, and, surprisingly (considering his natural reticence), John Flamsteed, had a very strong influence on their contemporaries.

We have already seen in general terms that personal contact was influential in beginning the career of many astronomers. What formal education could do for them can be inferred from available contemporary sources. Up to the end of the fifteenth century, schools and universities taught the subjects demanded by the social structure. Noblemen's sons were taught at home by other members of the family or private tutors, as tutoring was one of the jobs that could be done by the clerks of the Church. This sort of learning through intimate relationships continued to be important, as much for the informal interests that could be handed down as for the material itself.

Schools, however, increasingly provided the first stage of the education of future clerics. In the Middle Ages, schools were mostly attached to ecclesiastical institutions. Eton, for example, began as a chantry of a college of priests. The boys were involved in all the prayers of the Church's day with lessons distributed between. The lessons involved reading the psalter and the book of prayer in Latin. Boys memorised rules of grammar by heart from traditional readers. Very few books were available, and there was correspondingly little departure from the traditional requirements.

The universities continued the process that the schools began. Grammar was studied thoroughly so that graduates would be able to teach it. The texts to be read were in scholastic theology, an amalgamation of pagan philosophers, mainly of course Aristotle, and Christian teachings. Such reforms as there were in the universities tended to mean nothing more than better methods of teaching the old subjects. This is not surprising, as movements for reform were motivated by the need for well-grounded teachers, or priests whose sermons could better expound the main tenets of Aristotle and St Thomas.

Outside the Church, lawyers formed the only large group with an interest in education. Lawyers naturally had their own educational requirements. For example, they needed to study legal French and the history of canon and Roman law. The best places to study law were Padua, Bologna and Ferrara, where the standards were high and were certainly not matched by Oxford and Cambridge. Returning students emphasised that improvements were needed in England. The Inns of Court, therefore, were centres for men returning from the continent with enthusiasm for change and for the new humanism of the Italian Renaissance. While this did not necessarily imply any interest in science, it did imply opposition to the purely scholastic studies of the Middle Ages. On the positive side of the balance, it implied an interest in Greek texts studied in the original, and it therefore opened the way back beyond Aristotle to the works of scientists such as Hipparchus and Aristarchus whose astronomy was fruitful both for its own sake and for the new avenues that it indicated for the future.

With the dissolution of the monasteries came a move away from the authority of the Church in education. Useful studies such as astronomy were still not taught in many schools but the rigid scholasticism of the Church schools was no longer universal. Many former monastic schools were handed over to the Church of England, certainly, but new schools being endowed were equally often put under the care of a borough or a merchant company.

New schools were founded for the children of the poor, particularly in the reigns of Edward and Elizabeth. Christ's Hospital, under the management of the City of London, became one of the most important foundations for the teaching of mathematics in the late seventeenth century. The opening of education to boys of all social classes was certainly significant in allowing the wide spectrum of social backgrounds from which seventeenth-century astronomers were drawn. The pool of ability became much larger.

The expansion of the number of schools in existence was both a cause and a symptom of the growing desire among tradesmen and yeomen to use education as a means of improving their sons' social position. In some cases, schools were established in response to local demand. For example, at Mansfield and Fordmanchester in 1561 schools were founded because of a petition from the townsmen.

By the end of the seventeenth century there were many new social groups who might expect, as a matter of course, that their sons would go to school. There were, for example, the sub-professional groups such as solicitors and apothecaries, the skilled craftsmen, such as makers of scientific instruments, and, probably most important for the practice of astronomy, that entirely new phenomenon, the married clergy.

Many boys were being sent to school, but when they arrived, how likely were they to be prepared or encouraged for any scientific studies? At the beginning of the seventeenth century Bacon complained of the unfruitfulness of the 'philosophy of words' that was communicated in the schools. John Wallis, in 1631, learnt no mathematics or arithmetic at school, as he points out in his autobiography. As the century went by, however, some changes began to appear. Mathematics remained a subdivision of logic but some arithmetic was taught. Jonas Moore published his *Arithmetic* as a text book, and schools taught some mathematics, using other texts such as Robert Recorde's *Grounde of Artes* and Thomas Hylles' *The Art of Vulgar Arithmetick*. Yet, by the end of the seventeenth century there was still no more than the most elementary mathematics in the majority of schools. By the mid-eighteenth century Adam Smith complained that the schools still did not teach children modern geometry or mechanics.

In spite of a general expansion of the availability of education for the middle classes, for the children of the poorest classes schools were often still out of reach. Although grammar schools were nominally free, this meant only that there were no actual tuition fees. Parents still had to provide books, writing materials and wax candles. Moreover, in many schools there was an entrance fee. The masters' salaries were low. As a result they were inclined to exact 'gifts' and offerings from their pupils. Some of these offerings had become traditional at certain seasons. On Shrove Tuesday, at the cock-fighting sessions which continued through the sixteenth century, 'gifts' were demanded for the master and could vary from 6d to 2s 6d. This amount could be impossible for some families to pay. A poor labourer earning 3d a day needed his sons to work and earn money for the family as soon as possible. Grammar school, with its attendant expenses, would be out of the question.

Even for these groups the picture was not totally black. In addition to the grammar schools, the small country schools provided some elementary teaching. There would be few extra expenses in these schools, and even the poorest children might be able to attend for a year or two. Here they would be taught to read and write. During the sixteenth and seventeenth centuries, elementary subjects would be more and more frequently taught in English. Yet, even in these schools, some Latin grammar was still taught and a few texts were prescribed for reading. Such schools might provide a basic literacy for the 'petties' as the youngest children were called. Isaac Newton, for example, began his education at Woolesthorpe Dame School. From there, he progressed to the grammar school at Grantham, where he lodged with the apothecary, Mr Clark. He showed little interest in his studies and spent his spare time 'knocking and hammering philosophically' (Westfall 1981). He made water clocks, a model windmill and, at the age of nine, began on sundials. He made lanterns for kites so that, as they flew, they would shine in the dark, and he measured the force of the wind by comparing the distance he could jump with the wind behind him on different days. Yet Newton made no effort in his studies. It was a fight with a boy who had punched him in the stomach that seems to have spurred Newton on academically. He challenged the boy, beat him and banged his head against the church wall. After this he determined to beat him in every other way, and soon rose to be first in the school. The price he paid for this was guilt. In 1662 he listed among his sins 'beating Arthur Stower' (Westfall 1981).

Both in the grammar schools and country schools the preponderance of Puritans among the masters ensured a questioning pragmatic approach in the pupils. Puritan writers on education constantly emphasised that the purpose of education was to achieve both knowledge and understanding. Knowledge alone was worthless. Puritan masters were likely to set exercises to check understanding, not just retention, by asking children to make notes on a sermon and discuss it afterwards in school. Similarly, Puritan ideology was likely to de-emphasise Latin because of its close association with Catholicism. Boys who attended schools with masters of this sort were likely to be open to new ideas and well disposed towards experimental method.

Thus, in various different ways the foundations for astronomy were laid. The grammar schools and apprenticeship system can be seen to have played a larger part than might have been expected, although the public schools had individual teachers who encouraged scientific interests. Historians have looked closely at the part played by the universities in developing scientists and the conclusions reached have been varied. Hill (1965) sees science encouraged in the universities only during the period of Puritan domination, 1645–60. Shapiro (1969) argues that there was continuous growth and internourishment, with the most brilliant Gresham professors taking their ideas from London to Oxford and Cambridge. (Gresham College was founded by a private bequest in the reign of Elizabeth for the further education of adults in the City of London. There was a professor of astronomy, among other subjects.) Kearney (1970) believed that Oxford and Cambridge were uniformly resistant to change and inimical to experimental science: 'Those sciences with a utilitarian implication developed outside the universities.'

More recently Frank (1973), on the other hand, has argued that Oxford and Cambridge played a much more important part in science than has been recognised. University-educated men prevail among those with achievements in seventeenth-century science, according to the *Dictionary of National Biography*, out of all proportion to their number in the population at large. In a narrower field, of scientists in the *Dictionary of Scientific Biography*, 75% had been to university or college. Moreover, the achievers were men who stayed long enough to take degrees. Of all undergraduates at Oxford betwen 1560 and 1680 only 42% stayed to take a bachelor's degree, and only 24% took master's degrees. But Frank asserts that in his group of scientists the vast majority had higher degrees. Frank argues that the length of time spent to acquire a higher degree was enough to allow scholars to acquire habits of thought and research techniques that must have helped their work. Moreover, and perhaps most importantly, they met each other and became involved in research programmes that began to receive the attention of the community.

When we look at men specifically or mainly involved in astronomy, we find that a high proportion of those whose names became famous had been at university: those who had been at Cambridge slightly outnumber those who had been at Oxford (see table 3.1). The Gregorys are a reminder of the existence of the University of St Andrews.

Most of the debate about the part played by universities has centred on

Table 3.1 Undergraduate members of the universities with a major interest in astronomy.

Oxford	Cambridge
Boyle	Cotes
Caswell	Flamsteed
Derham	Harris
Desaguliers	Newton
Halley	Oughtred
Hooke	Pell
Pound	Pope
Wilkins	Rooke
Wren	Brook Taylor
	Wallis
	Ward
	Whiston

what might be learned there as part of the curriculum. Frank (1973) points out that the university statutes of 1636 at Oxford were extremely specific about forms of disputations, academic dress, etc, but were very vague about subject matter. This feature of the university regulations, both before and after 1636, meant that everything depended on the interest, energy and teaching ability of individual dons. John Wallis in 1632 at Emmanuel College found no-one to teach him mathematics; in fact 'I do not know of any two, perhaps not any who had more of mathematics than I (if so much) which was then but little'. Mathematics and astronomy were simply not 'in fashion'. The reason that he gives for this is social. 'For Mathematics (at that time with us) were scarce looked upon as Academical studies but rather Mechanical: as the business of Traders, Merchants, Seamen, Carpenters, Surveyors of land or the like.'

Wallis' comments bear out Kearney's (1970) view for the early part of the century. As the new philosophy gained ground in London and, later, at the Royal Society, there is no doubt that individuals at Oxford and Cambridge both practised and taught experimental science. Isaac Barrow in 1636, for example, was able to say that undergraduates in Oxford were proficient in chemistry and botany, two 'modern' experimental sciences. Seth Ward pointed out that if a professor wished he could add Copernicus to Aristotle and Ptolemy.

The position at Cambridge was similar. A helpful statute of 1549 allowed more space in the curriculum to be given to mathematics (which included astronomy) and less to grammar. As the seventeenth century progressed, more mathematics, physics and astronomy were taught. Moreover, activity at Oxford in the mid-century involved modern astronomy, and work centred on refining a satisfactory heliocentric system. Through the work of Rooke and Ward, who observed the comet of 1652 and then lectured on it, undergraduates could have had contact with observational astronomy. There were experiments in instrument-making, too. Wilkins and Ward, for example, tried to build an 80-foot telescope, with which to see the whole of the Moon at once. At Cambridge, astronomical observations were being made as early as the 1630s by such astronomers as Samuel Foster, John Palmer and Jeremiah Horrocks, all of whom accepted, with Copernicus and Kepler, that the planets rotate round the Sun.

All this activity has to be inferred from the work and writings of individual astronomers. The official statutes of the seventeenth century give little indication of what the curriculum contained. They generally establish minimum requirements and leave individuals to fill out the detail of the teaching. Surviving notebooks of undergraduates at Oxford in the 1680s show variation from college to college in the kind of books read, a clear indication that individual interests among the teachers dominated the curriculum. For instance, Kearney (1970) mentions the notebook of Charles King who, in 1682, studied optics, a highly practical subject which must have brought him into contact with new ideas in physics and, probably, astronomy.

The picture presented by the subject matter of published lectures shows similar variation. The statute establishing the Savilian Professorship of Astronomy at Oxford in 1619 required the incumbent to lecture on Ptolemy's *Almagest*, and also to lecture on 'Copernicus, Geber and other recent astronomers'. The emphasis that various professors and lecturers gave to the traditional thought of Aristotle compared with the new ideas of Copernicus and later Kepler, Galileo and the 'new astronomers' varied according to the individual. The first Savilian Professor, John Bainbridge, published a description of the comet of 1618, in which he showed that his thought was still dominated by Aristotle. Though he toyed with Aristarchus' heliocentric explanation for the movement of comets, he dismissed it, in the end, as too dangerous to entertain. When Henry Briggs became Savilian Professor, on the other hand, he taught Copernicanism in his lectures between 1620 and 1630. All that can be seen here is a wavering trend towards discussion of modern ideas in astronomy.

Later, the subjects of lectures became more clearly connected with the developments of contemporary science. In 1684 Newton gave a series of lectures at Cambridge, 'De Motu Corporum', which were published in 1686 as the first ten sections of the *Principia*. Undergraduates then had an opportunity to hear of the most exciting and most recent developments in astronomy and physics. The idea that university lectures might contain new work rather than the oldest of the old continued to grow. Whiston, Lucasian Professor at Cambridge from 1702, lectured on his own theories of astronomy and cosmology, putting forward current controversies, such as his own belief that the Earth's surface had been largely formed by collision with a comet. Whiston and Cotes together gave the first lectures on physics to be accompanied by demonstrations of experiments. Many lecturers hoped to publish their lectures as Whiston very successfully did, and therefore became more and more likely to put their latest and best work into them.

Cambridge welcomed experimental science so wholeheartedly that, in 1708, Roger Cotes was attempting to set up an observatory over the gatehouse of Trinity College. Stephen Gray, the dyer from Canterbury, travelled to Cambridge to act as Cotes' assistant and, in a letter to Flamsteed, gave an impression of what university astronomy could look like to an outsider. Gray was extremely scornful of what he found there and stayed only a few months. He never says exactly what happened, but he writes to Flamsteed from Canterbury on his return as though Flamsteed had advised him not to go in the first place. 'I had better have taken your advice which was more agreeable to my own inclination had I not been persuaded by the solicitations of my friends but we little thought or suspected such men could have been soe mercenary as I finde they are' (6 September 1708). Gray felt that he had been lured to Cambridge under false pretences, and certainly did not see it as having much to offer him as an already practising, self-taught astronomer. His scorn for the observatory is profound: 'I saw nothing there that might deserve your notice there was indeed that which they called their Observatory for noe other Reason that I could perceive than that some time or other they intend to make it soe' (6 September 1708). He says that they have at last obtained a sextant with which they plan to do great things, but Gray doesn't believe a word of it. They propose, for example, to make a catalogue of fixed stars. As Gray knows that Flamsteed's life work is just such a catalogue, he is totally confident that no-one else can do it as well as Flamsteed.

This letter shows the difficulty of reaching an objective conclusion about the value of work done in the universities. Certainly Gray's letter was strongly coloured by personal disappointment and partisan feeling. Nevertheless, it was Flamsteed's *Historia Coelestis*, made in his government-sponsored Royal Observatory, that became for some time the standard catalogue of fixed stars.

The broad picture suggests that individuals would be likely to make use of the schools and the universities. This is certainly the case, but there are also other channels through which astronomy, mathematics and, particularly, the practical parts of the science were handed on.

Of Flamsteed's colleagues and assistants the vast majority were self-taught in mathematics and astronomy. Of those who went to a grammar or public school, only the ones who happened to have a master who both was a capable teacher and had an interest in scientific subjects were likely to be helped greatly. Roger Cotes and Edmond Halley, for example, were fortunate enough to attend St Paul's School when Dr Gale was Master. As a result, Aubrey tells us that Halley grew so well versed in the celestial globes that 'if a star were misplaced in the globe he would presently find it'. Moreover, 'at 16 he could make a dyall and then he saide, he thought himself a brave fellow.'

Another public school offering a high standard of traditional learning to its pupils was Westminster, under Dr Busby. Christopher Wren went there at the age of nine, but had to flee to the country as his orthodox Anglican family fared disastrously in the Civil War. Robert Hooke was more fortunate. He lived in Dr Busby's house. In his time at the school he learnt Greek, Latin, Hebrew and oriental languages. He is said to have mastered six books of Euclid in one week, and certainly devoured all the mathematical knowledge that was in the London air at the time. Like Newton, he pursued his childhood interest in invention, but, unlike Newton, he seems to have been encouraged in it at school. Aubrey gives him credit for having invented '30 several ways of flying'. Even if not all these were wholly successful, the endeavour shows an interest in the laws of physics, and interesting failures make good teachers.

On the whole, however, the grammar schools and public schools alike, by the turn of the century, provided little more than a good working knowledge of Latin, which was still needed by a scientist, since much European correspondence had to be carried on in Latin. Hevelius, the Polish astronomer with no English, complained that the Royal Society was cutting itself off from Europe by publishing many books and papers in English. Most of Flamsteed's assistants and correspondents seem to have been at least competent in Latin and, therefore, probably went to a grammar school. Abraham Sharp, for example, one of Flamsteed's most prolific correspondents, was sent to Bradford Grammar School. Here he may have studied writing and accounts, because a little while after he left he set up a school of his own to teach these subjects. In spite of the high reputation of Bradford Grammar School at the time, it was no more ready to provide its pupils with the essentials of the mathematics that would be needed for astronomy than its contemporaries. Sharp began to study mathematics himself only after he had left. From the available Latin text books, perhaps, but more probably from masters with an individual interest, some sort of attraction to mathematics could still emerge strongly enough to make the youth, after he left school, wholly dissatisfied with an apprenticeship. He was apprenticed to William Shaw, a mercer in the city of York, but he left before the end of his eight-year term. Not being welcome at home after wasting his father's £20 premium, he went immediately to Liverpool to teach accounts and to study navigation and mathematics as best he could.

The education of many of Flamsteed's assistants is difficult to assess. Stephen Gray is one of the more obscure. We do not know for certain where Gray went to school, and this is true of many other assistants and correspondents. In Gray's case there are two schools that he could have attended. In Canterbury, at the time, the King's School was providing education for those who could pay. If Gray's father had been doing sufficiently well in his dyeing business he might have managed to pay the ten shillings a quarter that was the fee for commoners. The school archivist has searched the records and found none of the Gray(e) family listed as scholars. No lists of commoners remain, however, but there are mentions of some of the boys, showing that their backgrounds were socially varied. One was a 'widowe's son', another a 'venturer's son'. It is just possible, therefore, that Gray received his knowledge of Latin at the King's School. On the other hand, he, like his brothers, remained in Canterbury in trade, whereas an education at the King's School might have opened other possibilities. Histories of the King's School make clear that the teaching was still soundly classical (Woodruff and Cope 1968). Grammar and verse composition were the touchstones of a boy's achievement. Mathematics did not appear on the curriculum. Even if Gray's father had managed to find the money for the King's School, he would still not have been providing his son with all the tools of a future astronomer.

The other school in the city was the Poor Priest's Hospital. The building still stands in Stour Street, very close to the Gray family home. The hospital was established by Simon Langton and provided education for the sons of the poor. No lists of pupils from the period are known, and we have little indication of what the curriculum might have been. Possibly the practical needs of poorer boys were taken into account by the masters in teaching them some useful branches of mathematics. John Locke urged that gentlemen should learn accounts above all. Tradesmen's sons needed practical arithmetic even more than gentlemen's sons for their probable futures. Certainly, as a pupil at the Poor Priest's Hospital, Stephen Gray would have learned the Latin that he demonstrated later in his ability to read and discuss authors on astronomy who nearly all wrote in Latin. Gray's interest in astronomy itself is not likely to have developed at this school, and is more likely to have arisen through the friendship with Henry Hunt mentioned in chapter 2.

Flamsteed's other assistants and correspondents rarely mention their schooldays. Some of their origins can be guessed from chance remarks of Flamsteed, but rarely the detail of their schooling. He merely mentions that Abraham Ryley, 'a very ingenious young man . . . lives at Greenwich and is ready att numbers to calculate the Moon's visible places'. Although most seem to have come to Flamsteed already competent in calculating, whether they learnt it at a school, from a formal apprenticeship or from a relative, we can only guess.

Although some prominent astronomers had no father alive to teach them, the part played by fathers and all sorts of other relatives in teaching young men mathematics is considerable. For example, the fathers of Samuel Molyneux, Oughtred, Ward, Wilkins and Whiston all taught their sons mathematics, and were at least partly the instigators of interest in astronomy. Some of the fathers had come by their knowledge through a trade or business interest.

William Oughtred's father was the butler at Eton, and no doubt had a practical interest in mathematics. John Wilkins' father was a goldsmith who could teach his son mechanical skills as well. Seth Ward's father was an attorney and John Wallis and William Whiston both had ecclesiastical fathers. In Whiston's case, the help he received from his father was returned in many ways as the old man grew blind, deaf and lame and needed the young man's assistance. John Wallis said that he had learned mathematics from his younger brother in a vacation. The younger brother was learning mathematics for a trade. John was intrigued by the intellectual neatness of the subject. He asked his brother to teach him what he knew. That was 'my first insight into mathematics and all the teaching I had', consisting of common arithmetic, addition, subtraction, multiplication, division, the rule of three, etc. David Gregory learned enough to become professor of astronomy at St Andrews by studying the papers of his uncle James Gregory, who had been professor of mathematics at St Andrews before him.

Although direct transmisson of knowledge within families was a frequent form of education, it could hardly provide the source of navigators and astronomers that a mercantile, seafaring country required. A far better solution would be to make use of the schools, or to provide new ones. Mathematical schools opened everywhere. In London, by 1700, there was plenty of choice among individuals claiming to teach mathematics and navigation. The trouble was that very few of the teachers were competent either as mathematicians or as teachers. The only quality control came from the complaints of dissatisfied customers, particularly from seamen.

As education involved trade, it became a political issue and could not be left to academics—in some cases, however, politicians and academics were in agreement.

Charles II, motivated by the demand of his sailors and merchants for economic expansion and mastery at sea, agreed to set up a department at Christ's Hospital to teach mathematics and navigation to the boys. In 1677, Flamsteed, as Astronomer Royal, was asked to draw up a syllabus. The archives at the Royal Greenwich Observatory contain Flamsteed's original design, and from it we can see what a practising astronomer considered desirable as a training in the astronomy used in navigation.

Flamsteed's curriculum is designed for boys who have already mastered 'vulgar arithmetic'. He shows considerable understanding of the problems of teaching and an awareness of the psychology of learners. He suggests that the equipment needed by the master should include 'small rewards to encourage the boys that are found to do best at their examinations once a quarter'. The course is not to be theoretical, but thoroughly practical. Thus, the master will also need the most advanced equipment available: 'maps, charts, compasses, a quadrant, a good small pendulum movement, telescope glasses of 8 and 16 feet with tubes and a micrometer'.

When listing the subjects to be included in the curriculum, Flamsteed shows how far education has moved since the Middle Ages. Puritan methods have led to a consideration of the learner. The boys are to be taught 'Maine Trigonometry' but not by rote: the master 'shall shew them the ground and reason of each resolution and on what Geometrical theorem it depends, that they may be able to raise their rules or canons for themselves and not after the manner of common Seamen depend on their books for Rules without knowing the reason of them and their application to navigation'. The course provides the students with theoretical and practical knowledge, not only of the astronomy and mathematics needed for navigation, but also, specifically, for surveying and gunnery.

Flamsteed had done a useful job in drawing up a syllabus. He found that his responsibilities did not end there. One day, on his doorstep in Greenwich, there arrived some masters from Christ's Hospital, bringing with them two boys. They told Flamsteed without preamble that the King had decreed that two boys at a time were to be sent to Greenwich for instruction from Flamsteed. The masters were very offhand and quite unconcerned about Flamsteed's sense of anxiety over pressure from his astronomical work. In addition, they seemed hardly to be listening when Flamsteed asked about the arrangements to pay him; he wrote gloomily to Jonas Moore: 'the masters when they were here last took so little notice of any such thought that I know not whether I shall have so much as theire thanks for my paines.' His one hope is that 'they will thinke of a suitable recompense ere they change these for two others otherwayes I must desire to be excused the trouble of them since you know very well I have work of another nature under my hands'. Not, he points out, that he objects to the boys themselves. The first two at any rate are 'prompt and ingenious' and evidently the master, Peter Perkins, had taken a great deal of trouble with them. He had given them a good grounding but had not had time to teach them trigonometry.

Flamsteed resigned himself to teaching these two boys and subsequently many others that were sent both from the school and privately through the recommendations of friends and professional acquaintances. Filed near the letter to Jonas Moore is a letter that Flamsteed received some years later, in which a Mr Pratt is recommended as a private pupil for instruction in mathematics. He is being sent to Flamsteed by three employees of the

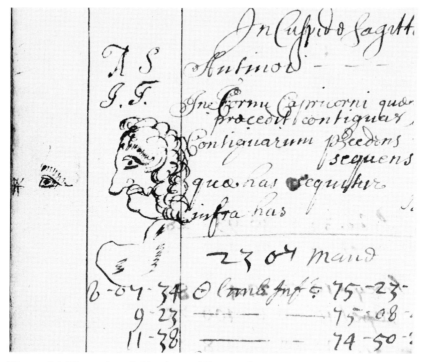

Figure 3.1 The work probably of one of Flamsteed's pupils doodling in the margin of an observing notebook. Royal Greenwich Observatory.

Office of the Ordnance where Jonas Moore was Master. This sort of private referral was common.

The story that these and similar letters reveal is that Flamsteed was finding his salary of £100 per annum inadequate, in spite of the stipend from his living at Burstow and the inheritance from his father. He could not afford to dismiss the possibility of fees from tutoring. When he sat down as an old man in 1709 to list the names of his former pupils, he was able to remember about 130 youths who had passed through the observatory. Of these, about 20 are listed as captain or lieutenant. Of the others, many are listed as 'at sea' or 'died at sea'. Some are working in the Office of the Ordnance, and John Wootan is keeper of the customs at Galloway. The majority apparently came to Flamsteed to learn astronomy for practical purposes, not for the intellectual pursuit in itself. There were some, however, who, while they were practical men, were not obliged to make use of astronomy to earn a living. Lord Guildford and Lord Dartford both came. Lord Archibald Hamilton, 'sone to ye duke', is also listed, but he put his learning to practical use and went to sea as a captain. His servant

John Shields is listed with him. If John Shields went there to learn navigation and go to sea with his master, this is an interesting example of science bringing together men from very different social backgrounds in the seventeenth century.

A close look at the list reveals another aspect of the working of the scientific community. Many of the names that appear on the list are men who later became Flamsteed's correspondents and assistants. Some of the boys who came to learn stayed on as full-time servants or assistants like Samuel Clowes and Isaac Woolferman; both learnt their astronomy this way. Others became occasional correspondents, writing from all over the world. The fact that men like George Slingsby began as boys under Flamsteed as master explains the very humble and reverential tone adopted later in their correspondence with him.

Flamsteed's success as a teacher of navigation and astronomy is another indication of the lack of opportunity elsewhere for learning these subjects adequately. Part of the problem towards the end of the seventeenth century was that the vitality of interest in modern science had led to many new developments. A result of this was that the average 'professor' or 'teacher of the mathematics', many of whom were hanging up their signs, especially in London, was incompetent to teach the subjects that he professed. Certainly the demand for mathematics and navigation was great. Improved instruments and methods of navigation were being developed, but would be useless to innumerate seamen. Even finding a method for calculating the longitude accurately at sea would lose a great deal of its economic value if sea captains and sailors were not capable of making use of it. The large number of teachers offering their services by the end of the seventeenth century is itself evidence of the demand. These teachers, however, presented a motley collection of qualifications. As there was no control, anyone could set up as a teacher if he could find pupils. Flamsteed's popularity down at Greenwich says something for his reputation, and possibly also for his teaching ability. Universities were still defining a teacher as one who could teach Latin syntax and accidence. A teacher with both practical and theoretical knowledge was rare. The difficulty of finding a suitable candidate to fill the post of mathematical master at Christ's Hospital is a clear indication of the extent of the problem and shows the sorts of scientific background that could be found among astronomers and mathematicians. Forty boys were to pass from the ordinary Latin classes each year to the mathematical school. A suitable candidate for the mastership would have the usual Latin which could certainly not be abandoned. He would need the pure mathematics of the university man, but since the boys were destined for the sea and would, in fact, be examined by Trinity House in practical navigation, the master had to be able to instruct them in the practical requirements of the sea.

Dr John Pell was the first nominee for the post. He had apparently in many ways a very suitable background. Like Hooke, he had struggled through an education without a father. Like Hooke's, his father had been a divine, but had died when Pell was only five. According to Aubrey, the Reverend Pell bequeathed to his son an excellent library. Pell made good use of it and entered Trinity College, Cambridge, at the age of thirteen and a half, already able to read Latin, Greek and Hebrew. In spite of such a promising beginning, his career gave him no great satisfaction. He always worked very hard, straining every fibre to such an extent that he complained that problem-solving gave him diarrhoea. He published little, however, and he was rarely paid in full for the work that he did. Aubrey points out that 'he was a most shiftless man as to worldly affaires and his tenants and Relations cosin'd him of his profits'. When he died, he 'had not 6d in his purse'. Disappointment was the rule of his life. He died of a broken heart.

As a candidate for the mastership of Christ's Hospital, though, Pell looked very promising. He had spent some time as a school master in Sussex, but, more importantly, he had published a *Description and Use of the Quadrant* which Henry Briggs found to be of high quality. His interest in education was demonstrated by *An Idea of Mathematics*. This was a scheme for making mathematics more accessible to the public at large. Hooke published the idea in a collection of his own in 1650. The main innovation that Pell suggested was setting up a library containing not only mathematical books but also mathematical instruments, all to be under the care of a helpful librarian/educator. If this idea had been put into practice, it would itself have made a considerable difference to the facilities for mathematical self-education in London. But Pell's plans had a habit of remaining no more than plans.

An important qualification for the mastership that Pell could offer was a knowledge of navigation, at least from a theoretical standpoint. His voyaging had been limited to crossing the Channel, not an inconsiderable obstacle when the crossing from Dover to Calais could take 48 hours in stormy weather. Nevertheless, he could hardly claim experience of navigation. He had, however, studied the problem of the variation of the compass, and in 1635, while still teaching in the obscurity of Sussex, he sent to Mersenne in Paris a commentary on Gellibrand's discussion of the variations of the compass. This tract was discussed at the Royal Society, but never published (Taylor 1954).

Like so much else in Pell's life, the proposal to make him master at Christ's Hospital looked like a good idea, but came to nothing. After first accepting the offer, he later turned it down, and served, instead, on the Commission to study Henry Bond's proposals for finding longitude at sea.

The next candidate proposed was John Collins. His life and work were a demonstration of the openness of seventeenth-century science to nonacademics. As a youth, Collins was a pupil of William Marr, clerk to the kitchen of Prince Charles. Marr was able to teach Collins more than the basic arithmetic needed to become assistant clerk to the kitchen. Marr himself later became a craftsman, maker of sundials. He made new sundials for the royal gardens at Whitehall, and later still became a surveyor involved in remapping London after the Great Fire (Taylor 1954). Through Marr, Collins met and made friends of many of the London craftsmen instrument makers and mathematical practitioners. In spite of his low-paid jobs as clerk in various government offices, and his marriage to the daughter of the chief cook in the royal kitchen, which can have been of little help to him socially, Collins made himself a centre for correspondence on scientific matters. He linked many of the practitioners of science and mathematics, both in England and further afield. He was highly praised by contemporaries for furthering communication and suggesting new and useful avenues for research.

Collins had spent the Civil War years at sea, and could claim the advantage of practical experience. The governors therefore decided to overlook his lack of academic qualifications. Collins described his own attainments as 'mean'. He had the Latin of a schoolboy and no Greek at all. When he was offered the post of master he actually turned it down and urged that it be offered instead to his friend and correspondent, Michael Dary. Dary was another self-taught mathematician working as an excise officer and as gunner at the Tower. Working at the Tower, Dary later became acquainted with Jonas Moore, to whom in 1678 he eventually 'humbly submitted' his *Doctrine of Adfected Equations*. In 1673, however, he had just begun work at the Tower on Tuesday and Thursday mornings, standing watch every fourth night. In spite of Collins urging his ability in arithmetic and navigation, he was not seen as having sufficient status to be a suitable candidate, and the job was finally offered to John Leake, who, in fact, had less.

Leake was another of the group of craftsmen/mathematicians in London. He made sundials and was a friend of Jonas Moore, who respected his ability to the extent of asking for his criticism of the *Arithmetic* before it was sent to the press. Leake, like the others, had no academic standing, nor, as it turned out, any teaching ability. He found the management of forty boys so difficult that complaints were made about him, including suggestions as to how else the boys might be taught. To Hooke's great annoyance, one suggestion was that the boys should be taken along to his geometry lectures at Gresham College. He managed to avoid that fate. Instead, the governors appointed an assistant to help Leake to manage the boys. Jonas Moore still did not find the standard of teaching satisfactory and, therefore, as Hooke was determined not to have boys, he originated the idea of sending boys to Flamsteed. This alone was enough to make Hooke very unpopular with Flamsteed. In 1678, Flamsteed expressed his feelings about the Bluecoat boys more freely in a letter to Richard Towneley than he had to Jonas Moore: 'Since then I have been lumbered with more business than I am well able to look after because I complained not before. Sir Jonas has thought fit to add to my burdensome employment and now I have every three weeks a couple of Bluecoat boys sent down hither to learn the stars as he pretended. Though I am satisfied with another much meaner design, these I am forced to instruct daily with no little trouble and much hindrance to my other studies.' In spite of his own inconvenience, Flamsteed was a strong supporter of Christ's Hospital, and, in fact, he urged Samuel Pepys, a Governor of the School, to appoint scholars as masters and to further the cause of educated seamen. The success or failure of a voyage might then depend on sound knowledge more than on 'fate or fortune', as had too often been the case.

Isaac Newton's opinion in a letter to Pepys was similar. He made the cogent point that 'if instead of sending the Observations of seamen to able Mathematicians at land, the land would send able Mathematicians to Sea, it would signify much more to the improvement of Navigation and safety of Men's lives and estates on that element' (Taylor 1954). In spite of this general theoretical support for the teaching of mathematics, the task of finding suitable masters continued to perplex the governors. Peter Perkins was probably the most effective master appointed in the seventeenth century. He was not a university-educated man but was made a Fellow of the Royal Society for his work on magnetics. He became a friend of Halley and Flamsteed who tried to acquire his papers when he died. Unfortunately, his death came after he had held the post for only three years. Once again, debate erupted over appointing a traditional academic or a self-educated practitioner, and this time Jonas Moore was not present to advise Pepys, who was in charge of the appointment, because he himself had died just before. Pepys' own view was that sailors were not a good source of academic knowledge. All our attainments in mechanics and even of navigation have come not from 'Tarpaulins' but from the 'chambers and from the fire sides of thinking men within doors'. Pepys ignored the need to develop and test methods of navigation in the difficulties of a tossing, spray-soaked deck. Not surprisingly the choice was made for an academic: Edward Pagitt, who turned out to be highly unsuitable. Isaac Newton recommended him because he was a fellow of Trinity College, Cambridge, and had made some study of applied mathematics. He had no practical knowledge of navigation nor, once again and even more disastrously than John Leake, of boys, but his dissolute behaviour was the final straw that weighted the decision to dismiss him in 1695.

Samuel Newton, appointed next, was once again a non-university man. He kept a mathematical school at Wapping and showed sufficient interest in education to write a textbook on navigation. Unfortunately, he himself lacked an adequate education. He knew too little to be able to teach others and standards fell. In March 1709 he resigned or, as Flamsteed put it, 'was turned out for insufficiency'. At last a thoroughly satisfactory appointment was made. James Hodgson had been Flamsteed's assistant and had married his niece. He was an able mathematician and astronomer in his own right, and was elected Fellow of the Royal Society. He stayed at the school for nearly 50 years. In this way the eighteenth century began with the rising status of practical mathematics and astronomy reflected in the establishment of at least one efficient educational institution.

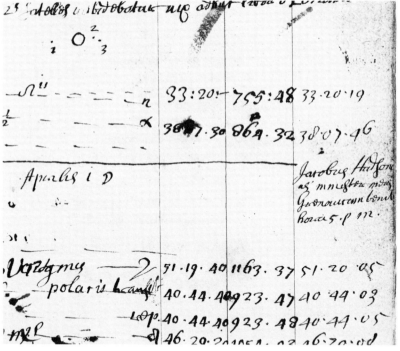

Figure 3.2 In the observing notes for 1 April 1695 this note appears: 'Jacobus Hodgsonus minister meus Grenovicum venit hora 5 pm [James Hodgson my assistant came to Greenwich at 5 PM].' Royal Greenwich Observatory.

During the struggle to establish a school, adult private pupils continued to flow past teachers like Flamsteed. In London, although there were many teachers of mathematics, knowledge of astronomy like Flamsteed's was impossible to find. He could teach directly from his own research like the best university teachers. But he was down at Greenwich. In London itself the Gresham College lectures, mostly given in English rather than Latin, provided some instruction in geometry and astronomy, but to a practising seaman they were inadequate. Henry Thomas, captain of the *Humber* at Spithead, wrote to ask Flamsteed for more information about his theoretical work, *The Doctrine of the Sphere.* Thomas explained that he needed help because of 'the insufficiency of some Professors to teach the mathematics in London'. Flamsteed was willing to teach casually on this sort of request, but he made lasting impressions on his long-term pupils.

John Witty, a former pupil of Flamsteed's, wrote to describe the problems he was encountering in his new job. In a letter of 15 September 1705, he says that he is determined to continue making astronomical observations and calculations if he can find even 'tolerable conveniency'. To continue may be difficult, but 'this will be an inducement to me rather to leave this place than to sacrifice all the knowledge I have had from you in that science'.

By November 1706, Witty's problems, whatever they were, had increased so much that he had given notice to his employer, Mr Wallop, and intended to return to Greenwich as soon as possible. He was toying with the idea of writing an introduction to astronomy which he saw 'wanting in a high degree', but had received some sort of discouragement: 'I perceive you are against my meddling with an introduction to Astronomy. I knew it would be a difficult task but it would be taking pains in a beautiful science.' He was full of admiration for Flamsteed's ability: 'if it were done by you it would be an inestimable treasure.' He hinted that if Flamsteed were not going to produce such a treasure, he might help Witty to produce something worthy of 'the dignity of the subject', but concluded that Flamsteed knew best and he would say no more.

Flamsteed, aware that he had not been very helpful to Witty so far, did the one thing that he could. He took him back into service as an assistant. Witty, apparently cheerfully accepting Flamsteed's embargo on writing a textbook, came back to Greenwich as another witness of the power of Flamsteed's authority and the importance of his practical help. This was not an isolated occurrence. Stephen Gray also acknowledged Flamsteed's authority and influence. He also toyed with the idea of writing a book but withdrew immediately from the project when he knew that Flamsteed was interested in doing something like it.

Flamsteed must have been a hard taskmaster. His anxiety for accuracy must have communicated itself from the first days when he taught a new and able boy like Abraham Ryley to 'calculate the Moon's place', and continued throughout the whole acquaintance. Sometimes Flamsteed had to teach by correspondence. Luke Leigh, a poor relative of Edmond Halley, learned to calculate at his home in Derbyshire. Writing to a mutual friend, the local apothecary, Flamsteed sent a message to tell Leigh 'to be very careful in taking out the Requisites for they are the foundation and an Error in them propagates through the whole work'. If ever the results were slow in coming back to Greenwich, a letter was sent to convey Flamsteed's displeasure, with the rather peremptory 'Pray let me hear from you'. Leigh suffered from a serious illness in 1705 and the flow dried up altogether. Flamsteed had to write to Bosseley the apothecary to ask about Leigh: 'I would employ him and therefore desire you to let me know how he does and wether his distemper doth not affect his head.'

Assistants on the spot were allowed some autonomy. John Witty made observations himself, and so, we can assume, did others. Once he had taught them, Flamsteed trusted their care, although he trusted no-one's accuracy entirely. All calculations for the *Historia* were done by at least two people providing a check on each other's accuracy. Even when opinions differed from Flamsteed's own, they were treated with respect. Nicholas Stephenson in 1675 'guessed his distances rather lesse than mine through the wideness of the sights. I thought the same but durst not adventure to diminish it.'

Judging by his finished work, the training provided by Flamsteed in practical astronomy was effective. He produced a number of observers and calculators who were highly skilled in practice, although not inclined to produce original or wide-ranging work of their own. The research programme to be pursued was determined entirely by Flamsteed and was always seen by all those involved as Flamsteed's work, not as a team project. Nevertheless, Flamsteed and his assistants are the only example from the period of collaboration on a large enough scale to produce a comprehensive, quantitative work. The existence of some external forms of training in mathematics enabled Flamsteed to teach his youths the particular skills that they needed in a reasonably short time. The universities produced other great astronomers or contributed to their production, but played little part in the work of Flamsteed and his group. An astronomer could learn what he needed in a variety of ways and inside or outside the educational institutions.

4

Personal lives

The personalities of those who chose to spend some of their time in astronomy are naturally varied. Certain qualities are more useful than others and might be expected to predominate. For example, the demands of research, then as now, ensure that anyone who succeeds will have intelligence and some strength of purpose. Independent thinking and creativity might be expected in those who initiate and plan research or put together theories. An interest in objects rather than people and an enjoyment of manipulating the symbols of mathematics would also be probable.

Among seventeenth-century astronomers, all these qualities are visible, and in many cases are quite outstanding. They also brought with them negative faces. Many of these men found difficulty in cooperating with others, or even in relating to people at all. Strength of purpose could show itself as stubborn and implacable hatred, and intelligence could be perverted into furthering quarrels and satisfying hostility.

To begin with the Astronomer Royal, Flamsteed seems to have found relationships with other people difficult and a struggle that was not often rewarding. He had, like Newton, the image of being very difficult. Among the other members of the astronomical community, personalities fall between the two extremes of solitariness and sociability, but many are close to the profile that Flamsteed exhibited. Astronomy was a solitary occupation at the time, although an assistant was desirable. On the other hand, it could occasionally lead to a sense of community among its practitioners. Nevertheless, a tendency to solitariness is one characteristic that can be seen in a large number of cases. For Flamsteed, some possible explanations can be found. Flamsteed had a childhood without a mother.

His father was left as a widower with John, aged eight, and two small girls. John was sent to school like other boys, but after the river swim that gave him rheumatic fever he must have been cut off from other children's games (perhaps his scholarly tastes would have done this, anyway). When he was removed from school because of his ill health, he chose as his own sources of enjoyment various mathematical texts, such as Sacrobosco's *Spheres* and Fale's *Art of Dialling.*

From an early age, Flamsteed showed a somewhat practical approach to friendship. He may not have wanted friends solely for what he could get from them, but he certainly cultivated the ones he thought useful: 'This year I also became acquainted with my friends Mr George Linacre and William Litchford. I have affected the friendship of the former because of his knowledge of the fixed stars, of the latter for his knowledge of the erratic and judgments' (Baily 1835). Whatever Flamsteed's motives, his mind was sufficiently lively to interest these men who taught him what they knew.

As a youth, Flamsteed also had a share of pride and some concern for what others thought of him. He bought a copy of the astronomical tables of Thomas Street 'because I would not be seen with Mr Gadbury's book lest I be thought astrological'. There was a softer side to him, perhaps, but he showed little of it in his letters and memoirs. What he said about his family rarely sounded affectionate. He mentioned that his sister was 'discreet or rather witty enough for her time', but gave no other indication of fondness for her. He coolly assessed his father's reasons for keeping him at home instead of sending him to university. There was a good deal of suppressed anger in the relationship, not only over the issue of university but also over Flamsteed's interest in astronomy: 'My studies were discountenanced by my father as much in the beginning as they have been since, but my natural inclination forced me to prosecute them through all impeding occurrences.'

Hardly surprisingly, Flamsteed in later life had difficulty with relationships. His personality and early experience must have been involved in the problems that he had with Halley and Newton. These quarrels, however, are only one side of Flamsteed's character. He had stable relationships with many others. Jonas Moore, for instance, must have admired Flamsteed's ability as one motive for getting to know him, but, presumably, he did not dislike him personally, as he extended an invitation that was not only generous but sounded warm: 'I rejoice that I may again hope to see you and do with all earnestness beg from you that whilst you stay at London you will make my house your abode. I have a quiet house, a room fitted for you and another for your servant, and I have a library and all things else at your command.' He added that Flamsteed would be free to choose his diet and hours of rest.

Moore's own rewards were to include the satisfaction of Flamsteed's conversation. His letters show a great eagerness to find someone who would talk about the subjects that so fascinated him. 'I have many things that I desire to impart to you.' Moore was not disappointed by Flamsteed's presence when he finally arrived in London. The two were together for a time at Moore's lodging in the Tower. This closeness seems to have enabled them to develop a friendship. If Flamsteed's personality had been as difficult as is often suggested, the relationship might have been considerably cooler by the time Flamsteed moved out to live on his own at the new Observatory at Greenwich.

Correspondence between the two men that spans the whole of their relationship until Moore's death in 1679 has survived, and shows something of how they felt about each other. During those years Flamsteed wrote in a warm tone that is unusual for him. He signed himself 'Your ever obliged and affectionate servant' and spoke of many planned visits to the Tower, a number of which were cancelled because of the dangers of the journey by water to someone of Flamsteed's poor health and weak constitution. He sometimes managed it and thanked Moore for 'my late kind entertainment'.

The relationship between patron and patronised is bound to be difficult, and Flamsteed and Moore had their problems. Most of these were resolved sufficiently well to keep the relationship friendly, but Flamsteed's letters do take on a querulous note at times. He complained that no-one understood how much work he had, and, to his surprise, Moore was involved in sending Bluecoat boys from Christ's Hospital. Flamsteed was pained that Moore should be capable of ignoring the much greater importance of the sky survey by interrupting it with the enforced teaching of little boys. Even worse was the affair of the wall quadrant (see chapter 6). Flamsteed managed to bring Moore round to his own point of view and the comment here reveals his somewhat irritating tendency to moral superiority. Moore had made the foolish mistake of not consulting Flamsteed, but in the end, 'Sir Jonas was sensible of his fault and that this was an instrument wholly use less' (Baily 1835). Nevertheless, Flamsteed continued writing in a confiding way about his health, or lack of it, and when Moore died there is perhaps true sorrow, not just regret at a lost source of income, in Flamsteed's words: 'My good friend Sir Jonas Moore died on August 27 1679'.

The weakness and susceptibility that prevented Flamsteed from travelling to London as often as he would have wished, in order to visit Moore, meant that he suffered from various illnesses for much of his life.

In addition to the rheumatic pains which he never completely lost, he suffered the agonies of 'the stone' and also recurrent bouts of malarial fever, the 'ague'. Perhaps the marshes round Greenwich were a breeding ground for malarial mosquitoes. Certainly, in the 1670s he developed a bad attack, and Jonas Moore's grandson, who was studying with him, caught it at the same time.

Flamsteed's account of the illness given in his journals and letters to Jonas Moore shows the frustrating way in which the fever came and went, and how it interfered with work. In October 1677, Flamsteed recorded the beginning of a quotidian ague (malaria in which the fever comes every four days). 'My headache is the most troublesome paine but I more fear a small but constant cholick in my bowels.' Five days later he was 'cured by Dr Tabor's medicine'. In February 1678 it began again, worse than ever, now coming every three days. This time he was too ill to work. 'I have been soe constantly ill that from last Monday was a sevennight I was not able to examine the clock.' The attacks settled into a predictable routine: 'This is my well day but I expect my fit about 8 o'clock at night' (5 March 1679). He shakes for about an hour and then is ill for twelve hours after. What could be done? Very little. He hopes that if the weather improves he may get better 'without the doctor'. He takes nothing for it 'but a little curd posset drunk to fetch the flegme of my stomak . . . I use a very slender and spare diet. My cordiall is a glass of sack and some mithridate which I find fits me best' (2 April 1678).

In April, Flamsteed's fits are still just as bad, although Moore's grandson is better. Flamsteed is sure that he, too, would recover if the wind would leave 'this unhealthful quarter'. Under his discomfort, Flamsteed suffers most from not being able to work. Seeing instruments unused and calculations not made troubles him 'as much as my dystemper'. By 30 April, though, the fever has gone: 'My ague I thank God is wholly departed but it has left me paines in my feet, legges and armes.'

This last remission continued until July when he noted in the margin of some work done on the 20th: 'began to be ill an horrible fever and hiccough.' This time he was driven to consult a doctor. In August he went to London to see Sir Charles Scarborough who 'prescribed to me and through God's blessing I was cured'. He returned to Greenwich on 22 August with the fever gone, 'but weak'.

This is the sort of experience that must have contributed to Flamsteed's rather austere view of life. His correspondent Stephen Gray had a similar reputation for touchiness. He became a success with his electrical experiments at the Royal Society and John Desaguliers wrote that no-one else dare work in the field or Gray would be so offended that he would probably refuse to do any more work. Thomson, in his history of the Royal Society, went on to say that Gray's temper was uncertain and he was 'by no means aimiable'. Like most other scientists, Gray was not prepared to let others take the credit for his work and he certainly had suffered from lack of recognition in his early days. He too suffered from ill health, however, and we may perhaps conclude that some of his humourless approach to life was caused by the continual pains in his back from lifting heavy weights in his dyeing business.

Gray quarrelled with a number of people in his later years. In this, he was certainly not unusual among astronomers. Flamsteed quarrelled with a number of his better known contemporaries, quite apart from his most famous quarrel with Newton, over the publication of results. At various times he quarrelled with Samuel Molyneux and David Gregory. His most lasting disputes were with Hooke and Halley. In both cases, the hostility seems to have arisen more from a clash of personalities than from any great provocation. The strength of the hatred between Flamsteed and Hooke can be seen as an indication of the misanthropic tendencies of both men, for whom dislike of colleagues was not unusual. From a psychiatrist's point of view, both men might be expected to have had such problems because both had grown up as isolated children without mothering and without the opportunity to form early trusting relationships.

Flamsteed was not the only astronomer to find Hooke's personality trying. But, again, there were factors which may at least partly explain the difficulties of Hooke's relationships. Hooke had to bring himself up from the age of thirteen with no close relative to give him any affection. While at Westminster School, he lived with the headmaster, Dr Busby, with whom he seems to have developed a closer contact than the usual master–pupil connection. Later, he created a few close relationships for himself. He took his niece Meg into his apartment in London, giving her a home and treating her with generosity. He also developed a relationship with his servant, Nell. Money was always a problem for Hooke, yet his diary records regular amounts given to Nell for dress lengths of silk and for petticoats: 'Gave Nell £3 for faradine made up'. His diary records the closeness of this relationship by the use of the symbol ♓ every time he made love to her. The entry for 13 August reads 'Nell went out yesterday and abroad all this day. I suppose today married.' On Thursday, 'Nell came home', and, on Friday, we find again '*Mane* [in the morning] ♓ Nell.' On Saturday, 31 August, 'Nell's husband returned. She lay with him that night and ever since.' As a result, on 29 September, he paid Nell 20s for her last quarter's work, and dismissed her. He still visited her occasionally, but her place was taken by a new maid, Dolly. Probably the relationship was mostly physical, with little spiritual satisfaction, but it must have helped a man who seems to have been very lonely, in spite of his almost daily visits to Toothe's or Jonathon's coffee houses, where his intellectual relationships flourished.

In other ways, too, relationships with women must have been of some benefit to a man who was regarded by most of his contemporaries as very peculiar in mind and extremely ugly in body. Their comments were often unfavourable, although Pepys said that Hooke 'is the most and promises the least of any man that ever I saw'. Remarks about his personality were usually negative. Richard Waller, who edited Hooke's works after his death, said 'his temper was melancholy, mistrustful and jealous'.

Without doubt, Hooke quarrelled with more people even than Flamsteed or Newton. He called his colleagues and associates a rich variety of rude names. For example Tompion, the instrument maker, is variously 'a slug', 'a clownish, churlish dog', and 'a rascall'. Oldenburg is 'a lying dog', and, later, 'treacherous and a villain'. With probable basis in fact, he accuses Oldenburg, who had worked for the diplomatic corps, of being 'a trafficker and intelligencer'. The assistant Henry Hunt often came in for anger and abuse and was on the verge of being fired, but such threats came to nothing. Hooke's anger was fiery but did not usually lead to malicious action.

Physical descriptions of Hooke gave a clear indication of one important reason for his bad temper. He was 'in person but despicable being crooked and low of stature and as he grew older more and more deformed. He was always very pale and lean and latterly nothing but skin and bone with a very meagre aspect' (Waller 1735). Clearly, for much of his life, like Flamsteed, Newton and Stephen Gray, Hooke was ill. He includes in his diary an account of a most wretched regime of drugs and medicines designed to induce purging of various sorts. He records with satisfaction the number of times he made himself vomit or purge with a particular concoction. Occasionally, a new medicine would offer hope of health. A dose of sal ammoniac led to a heartfelt 'in the new world with new medicine'. The new world lasted about two weeks. He seems to have suffered basically from acute sinus congestion which gave him severe headaches, and he also records giddiness, indigestion, vomiting, sleeplessness, worms, catarrh and nightmares. When he died in 1703 he was suffering from blindness and 'swelling of the legs'.

While they may not excuse an irritable temper, constant illness and discomfort certainly explain it in part. Hooke's anger was also aroused by his feeling that his work was not appreciated or acknowledged by his contemporaries, and in this, again, he has much in common with Flamsteed and Newton.

John Pell was another who suffered physically. In his case illness was caused by living in a low-lying parish in Cambridgeshire, and his illness may have contributed to the failure to finish so many promising projects. William Whiston was forced to give up lecturing and teaching at Cambridge because of ill health. He complained that all his life he was an invalid, and that ill health was caused by having worked hard in confining circumstances when, as a boy, he had to be the eyes, ears and feet of his father, the rector of the parish, who had become blind, deaf and lame.

Illness did not prevent him from writing but it certainly caused him inconvenience for much of his life.

Isaac Newton is not usually thought of as a sick man and undoubtedly his problems were psychological as much as physical. He had a temperament that was, in many ways, similar to that of Flamsteed. His childhood, too, was unhappy. His father died before he was born and his mother re-married when he was three. A stepfather who takes some of the attention of a child's mother away from him is hard enough to bear. Newton's stepfather, the Reverend Barnabas Smith, removed the boy's mother completely. He took her to live at North Witham Rectory, but the little boy of three was left at Woolsthorpe to live with his grandmother. Westfall (1981) points out that Newton never mentioned his grandmother with affection and her death was not mentioned at all. As for his stepfather, we can infer the state of Newton's emotions from his 'list of sins', where he confesses to 'Threatening my father and mother Smith to burne them and the house over them' (Westfall 1981). In his psychoanalysis of Newton, Manuel (1968) sees Newton's unhappy relationships with Flamsteed and Hooke, among others, as constant attempts throughout his adult life to express the anger that he could not express as a three-year-old against the man who ravished his mother. Whether or not we accept such an account of Newton's emotional problems, we have to accept that he suffered acute disturbances in his mental balance.

Newton would be the more surprising if he could be described as 'ordinary' in any sense. He deliberately cut himself off from correspondence during the 1670s, and Westfall (1981) points out that by 1676 he had immersed himself in theology and chemistry and saw almost no-one. He lived like a prototype of the absent-minded professor of the comic strips. Humphrey Newton, his servant (but no relation), relays many stories about his absent-mindedness. He would sit in hall for dinner but forget to eat. When he had friends in for a bottle of wine, a thought would occur to him and he would sit down at his papers, quite forgetting his friends. Indeed, this story is hard to believe only because he had so few friends. Most of the time, even among company, he sat silently, occupied with his own thoughts. In five years Humphrey saw him laugh only once. He had lent a copy of Euclid to an acquaintance who then asked what use it would be to him. Newton at this 'was very merry'. Most of the time, however, he was not very merry. He broke off his acquaintance with another Fellow, John Vigani, because of a 'loose story about a nun'.

In the 1690s, Newton became seriously deranged. His frenzy lasted eighteen months. During that time he showed signs of paranoia. He wrote, for instance, accusing John Locke of trying to embroil him 'with women', a fantastic and rather pathetic accusation, as he had never shown any interest in women, except, perhaps, for falling in love in a boyish way with the daughter of the apothecary where he lodged in Grantham in his school days. In contrast to Hooke and Halley, Newton is reputed to have died a virgin. Newton himself later explained his state of mind in the 1690s as being caused by sleeplessness. He had not slept an hour for a fortnight, 'nor a wink for 5 nights'. One of the popular modern theories is that Newton was suffering from mercury poisoning, induced during his chemical experiments by tasting compounds and by inhaling vapours. One of the chief symptoms of mercury poisoning is sleeplessness. The evidence is not conclusive, although one group of researchers claims to have analysed Newton's hair and found abnormally high deposits of mercury in it (Johnson and Wolbarsht 1979).

Mad or merely exhausted, Newton's personality had already shown itself to have peculiar quirks by the time he came into contact with Flamsteed. The two men were both passionately committed to activities which seem to an outsider to be complementary: one was making observations, the other moulding them into theories. A story of them both has Flamsteed once accusing Newton of doing nothing original, but merely using the ore that Flamsteed had dug up. To this Sir Isaac replied that if Flamsteed had dug up the ore, it was Newton who had made the gold ring. Such an anecdote shows the kind of mutual suspicion and one-upmanship that each felt over the work of the other.

The one major astronomer with whom Newton continued on friendly terms was Halley. The *Biographia Britannica* of 1757 describes Halley as possessing 'the qualifications necessary to obtain him the love of his equals. In the first place, he loved them; naturally of an ardent and glowing temper, he appeared animated in their presence with a generous warmth which the pleasure alone of seeing them seemed to inspire.' This is glowing praise, and the philanthropic image is not necessarily the view of all Halley's contemporaries. There were times when his lifestyle was the subject of some serious criticism, especially after he had been away to sea and acquired some seafaring habits. He was accused of drinking, swearing and seducing women, but he was not often accused of being an unfriendly or difficult colleague.

Halley's personality provides a contrast with Hooke and Newton, as well as Flamsteed, in almost every way. Surprisingly, although Flamsteed quarrelled with Halley, and so did Hooke, the relationship between Newton and Halley survived until Newton's death. It was not friendship, exactly. Halley was fourteen years younger than Newton and his tone was always one of humble submission and eagerness to be useful. In seeing the *Principia* through the press, Halley undoubtedly performed a great and very arduous service. In the first place, he understood the importance of the work, referring to it as a 'divine treatise' and one 'that all future ages will admire'. His appreciation contrasts with the incomprehension of contemporaries as a whole, summed up by Dr Babington watching Newton go by in the streets of Cambridge: 'There goes the man that writt a book that neither he nor anybody else understands' (Westfall 1981).

Since Halley's temper was so much more calm than Flamsteed's, he was able to accept and forgive Newton's bursts of rage. He remained precariously in the middle of the anger that passed between Newton and Hooke, but kept his eye on his intention of getting the work finished in a worthy form. All of this bears out the descriptions of Halley as 'sweet and affable', 'always ready to communicate' and 'disinterested' (*Biographia Britannica*).

Newton accepted Halley's services with a gracious acknowledgment in the preface but, as far as is known, no personal thanks. The two men remained on speaking terms in so far as Newton spoke to anybody. Occasionally he grew angry with Halley for talking 'ludicrously of religion' (Macpike 1937).

Flamsteed quarrelled with Halley on this subject, among many others, accusing him of being a libertine and an infidel. Yet Flamsteed and Halley had once been friendly. Halley, as a young man at Oxford, wrote to Flamsteed, asking for the honour of a correspondence 'considering how free and communicative a genius you expressed in your satisfactory answer to the request of my very good friend Mr Charles Boucher . . . I am a true honorer of your worth and a real well wisher to astronomy and all its followers' (10 March 1675). Flamsteed accepted the offer of communication and the relationship began well enough. But all Halley's enjoyment of human nature and his reverence for astronomers was not enough to preserve this good will.

Flamsteed, like Newton, was an austere, religious man, and Halley was widely thought to mock religion. In fact, when the Savilian Professorship of Astronomy became vacant on the death of Wallis, Halley considered himself a candidate. Flamsteed wrote: 'Dr Wallis is dead—Mr Halley expects his place—who now talks, swears and drinks brandy like a sea-captain' (18 December 1703). Aubrey gives independent testimony to Halley's libertinism when he says that Halley sailed off to St Helena on his ship with a married couple who had been childless for many years. However, when Halley had spent time with them on the island, surprisingly, 'she was brought to bed of a child'.

Nevertheless, what it was in Halley that Flamsteed came to dislike so strongly remains obscure. Perhaps the problem in Flamsteed's case was guilt. Newton (who was not entirely disinterested) accused Flamsteed of including in his lunar tables data stolen from Halley. He even claimed to have seen Halley's handwriting in Flamsteed's manuscripts. David Gregory, who had his own reasons for disliking Flamsteed, repeated this story, too. It would not be surprising if some issue of this kind had come between them. In any case, Flamsteed never understood Halley's lighthearted approach to life. 'I hate his ill manners, not the man: were he either honest, or but civil there is none in whose company I would rather desire to be.' From this we can infer that Halley had offended Flamsteed's dignity.

Another possible reason for the quarrel was a letter by Halley, published in the *Philosophical Transactions*, in which he criticised Flamsteed's tide table published three years earlier. Flamsteed had neglected the effect that a river estuary would have on the tides, and his tables were, therefore, inaccurate. He was not pleased to be told so. Some passing raillery that meant nothing to Halley might have been enough to sow resentment in Flamsteed's gloomy heart. By 1716 relations between them were so bad that, when he had to communicate with Flamsteed, the usually good-natured Halley wrote in a cold, unsympathetic tone: 'Sir, I am commanded by the President, Council and Fellows of the R. Society to put you in mind that you are in arrears to them a copy of your Astronomical Observations for the year 1714.' This was a case where incompatible temperaments proved far more important than the possible advantages to be derived from cooperation.

The ill feeling spilled over to others. Flamsteed's assistants, Sharp and Crosthwait, took Flamsteed's side. Long after Flamsteed's death, Crosthwait wrote to Sharp: 'Dr Halley, so I hear, lives in taverns. He is very infirm.' Crosthwait adds that the quarrel lives on in Halley's anger that Flamsteed's work has been published. He will, suggests Crosthwait, concentrate on finding mistakes in it (28 July 1722). This is interesting as an indicator of how ill feeling had spread to partisans.

The enjoyment of good company could be a serious drawback to an astronomer in other ways than by earning the disapproval of Flamsteed. Roger Cotes, the brilliant young Cambridge don who died at the age of 34, tried to observe a solar eclipse in 1715 but, according to Halley, was 'opprest by too much company so that though the heavens were very favourable yet he missed both the time of the beginning of the eclipse and that of total darkness' (*Phil. Trans.* **XXIX** 23). Cotes was generally well liked, but there was always, as with Pell, a sense that he could have achieved more than he did. 'Had Cotes lived,' said Newton, 'we might have known something.' As it was, in Cotes' busy life there had not been time. He had not closed his study door with sufficient determination.

The temperaments of men like Flamsteed, Hooke, Newton and Stephen Gray suggest that they lived solitary lives without much satisfying emotional involvement with other people, although Flamsteed did marry. The number of astronomers who never married seems high and might suggest a general tendency to put work before relationships. Current medical statistics indicate that married men are, on average, healthier and longer-lived than unmarried men. Not marrying might therefore be either a cause or an effect of the ill health and depression that so many astronomers suffered.

In the mid-seventeenth century the Puritan regime and Civil War had encouraged late marriages. The religious atmosphere encouraged self-restraint and prudence and, with them, a general belief that young people should not marry until they were well able to provide for themselves. By the early eighteenth century such attitudes were much less powerful. The strength of Puritanism was dissipating into a secular spirit and, in many cases, into deism. The practical considerations of the use and value of a relationship became more important as reason moved into the ascendant and morality became more pragmatic.

Marriage was in many ways an economic proposition for a young man. A labourer or an apprentice would expect to marry and make use of the labour of his wife and children. Labourers might marry fairly young, but in other social groups marriages were often left until, for example, the apprentice had become a journeyman and had inherited his father's tools or workshop. Professional men and gentlemen would often leave marriage until they, too, were established, which often meant waiting until the father died; but, if a girl came with a large dowry, the couple might marry sooner. For economic reasons, widows and older women were considered a prudent choice. Defoe advises considering 'the homeliest and eldest' of an array of sisters, as she might be most helpful in 'application and business' (Porter 1982). One result of such prudent late marriages was that in many cases both parents would die when their children were still young, not only the mothers, who were, in any case, likely to die on their childbed. Hooke, Pell and Petty all at an early age lost their fathers, and an even greater number lost their mothers.

Astronomers were as slow to marry as any other group in the population, and a high proportion never married at all. Of those who did marry, a few took wives who remained in the domestic sphere, ran their houses, and in some cases bore children. Halley was one of these. In April 1682, only three months after his return from a visit to the Continent, he married Mary Tooke, who was the daughter of an official working in the office of the Exchequer. She may have had some money, but does not seem to have been a major financial gain to him. Presumably he married for love, or, at least, a comfortable domestic life. The *Eloge* of Halley (Macpike 1937) tells us that Mary was 'an agreeable young Gentlewoman and a person of real merit, she was his only wife and with whom he lived very happily and in great agreement upwards of 55 years'.

The Halleys set up house in the village of Islington, outside London. At home, Halley began observing with his sextant and telescope, for which he had made himself a small observatory. He gradually became more involved with the Royal Society until, in 1686, he was appointed Clerk and Assistant to the Secretary. Many of the scientists of the day must have visited his house, but we have no evidence that his wife became involved in his work in any capacity other than hostess.

Women were not often involved in science at the time. When a woman tried to take an interest she was liable to be mocked. The visit of the Duchess of Newcastle to the Royal Society in 1667 had been greeted with some derision. Pepys did not like the eccentric way she dressed: 'The Duchesse hath been a good comely woman but her dress so antick and her deportment so ordinary that I do not like her at all.' She was shown many fine experiments, including a very 'rare' one of turning a piece of roasted beef into pure blood. She was full of admiration at the sight, but Pepys did not hear her say anything 'worth hearing'. Lack of any kind of disciplined education was a serious drawback for a woman who was confronted with an even more momentous task in self-education than a man, who would at least have been taught the classics. The general judgment of the Duchess was that she had an imagination full of wonderful fantasies and might have had a brilliant mind, if she had knowledge and discipline.

Such a woman was, of course, a rarity. For the most part, women attended to their houses, their children, and such work as was allotted to them. There were exceptions, however. The Polish astronomer, Hevelius, after the death of his first wife, married a young woman, Elizabethe, who had a good mind, and not only helped run the brewing business but also assisted Hevelius in his observations for many years. In England we find, surprisingly, that the sick, sad John Flamsteed was one of the few astronomers to marry, and that his wife, too, became involved in astronomy.

Flamsteed married very late in life, even for a period of late marriage. When he married Margaret Cooke in 1692, he was 46 and she was 40. At this age, Flamsteed was probably past his prime. Moreover, as the daughter of a lawyer and granddaughter of the former rector of Bustow, Margaret brought him only £50 per year—hardly a great improvement to the Observatory's hardpressed finances.

A close look at the Observatory records shows another possible reason for the marriage. Margaret Flamsteed showed both ability and interest in mathematics and astronomy. Data and calculations in the observing note books are in her hand. Moreover, she was no mere secretary or note taker. She was herself involved in experiments. A letter from Stephen Gray includes a message to Margaret: 'I have made some triall to grinde Microscopicall Glasses in the tooles you gave me they Performe very well and soe soon as I have Opertunity I will send Madame Flamsteed some of them' (26 December 1703).

William Derham also sent glasses for Mrs Flamsteed's microscope. Apparently, she had given him some glasses first, and he was returning the favour by sending some well-cut double convex lenses. Mrs Flamsteed was

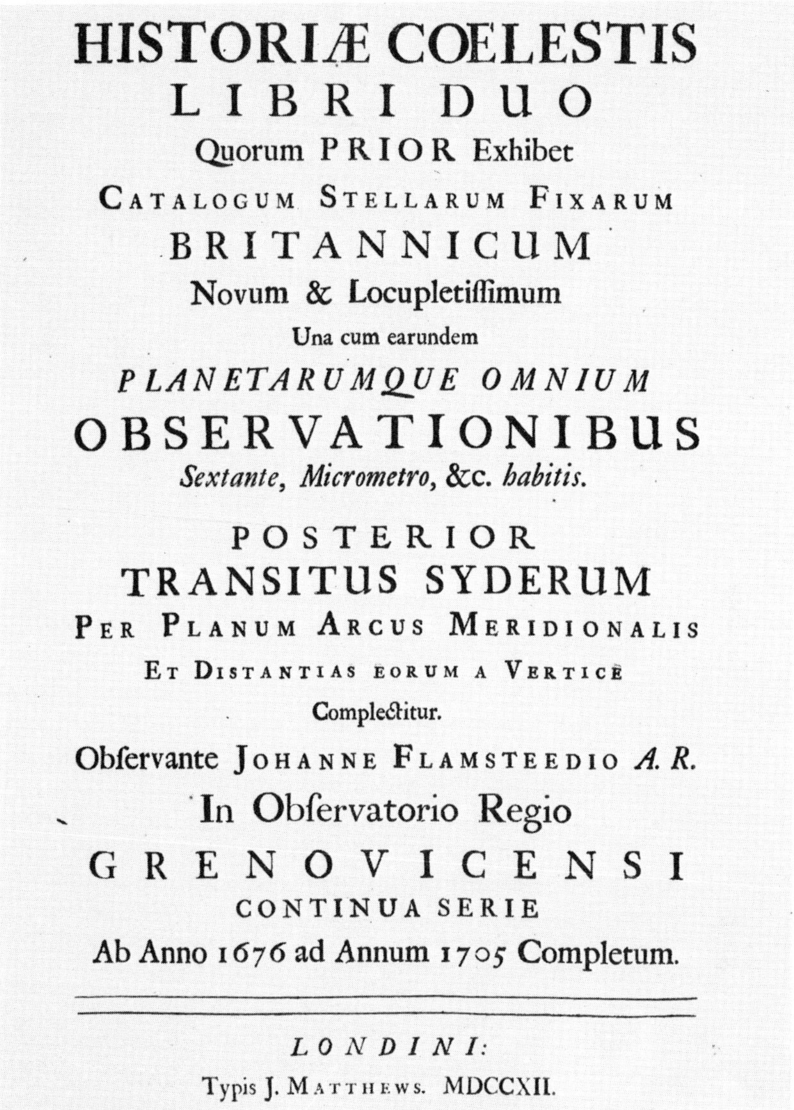

HISTORIÆ COELESTIS
LIBRI DUO
Quorum PRIOR Exhibet

CATALOGUM STELLARUM FIXARUM
BRITANNICUM

Novum & Locupletiffimum

Una cum earundem

PLANETARUMQUE OMNIUM

OBSERVATIONIBUS
Sextante, Micrometro, &c. habitis.

POSTERIOR
TRANSITUS SYDERUM

PER PLANUM ARCUS MERIDIONALIS

ET DISTANTIAS EORUM A VERTICE

Complectitur.

Obfervante JOHANNE FLAMSTEEDIO *A. R.*

In Obfervatorio Regio
GRENOVICENSI
CONTINUA SERIE

Ab Anno 1676 ad Annum 1705 Completum.

LONDINI:

Typis J. MATTHEWS. MDCCXII.

Figure 4.1 The title page of the 1712 edition of the *Historia Coelestis*. Flamsteed tried to find and destroy all the copies of this committee version of his work. The page reads: 'Two volumes of the *Historia Coelestis* of which the first shows the new and very thorough British Catalogue of fixed stars together with observations of all the planets made with the sextant, micrometer &c. The second includes transits of stars across the plane of the meridional arc and their distances from the vertex. The observer was John Flamsteed *A.R.* of the Royal Observatory, Greenwich, and the whole makes a continuous series from 1676 to 1705.' Royal Greenwich Observatory.

certainly interested in astronomy as well as microscopy and, besides mentions in letters to her husband, she received some letters of her own. Henry Stanyan, a former pupil of Flamsteed, sent her in 1706 a representation or drawing of a total eclipse of the Sun. His tone was a little patronising. He was sending it 'to divert you Madam for a moment', which suggests a low opinion of her powers of concentration. When he wrote to Flamsteed, his tone was even more patronising about ladies who were interested in astronomy: 'I had one Lady with me on the night of the Eclips that would have pleased Madam Flamsteed. She talked of nought but Apogeums and of Peregeums Arctick and Antarctick but very learnedly and was as well pleased with the Constellations of the heavens as if she had placed them there herself' (23 October 1706).

Mrs Flamsteed may have appealed to the astronomer as a congenial companion in his work. She was probably also efficient at maintaining a comfortable home. When John Witty finished studying with Flamsteed and left to work as chaplain in the household of Mr and Mrs Wallop at Hurstbourne, he promised to send to Mrs Flamsteed a journal of his life there. In one letter he apologises for not sending a minute catalogue of 'all the dishes we have had to supper and dinner all the last month' (5 September 1706). He excuses himself on the grounds that, since they had had 434 in that time, he could hardly be expected to list them. Mrs Flamsteed, as wife and mistress of the house, was evidently taking a professional interest in cooking and finding new recipes.

After Flamsteed's death, his wife showed loyalty to him, and perhaps her hard work to ensure that his results should be published as he had wished can be construed as a sign of affection as well as loyalty. Flamsteed had completely rejected the version of his star catalogue, the *Historia Coelestis*, that had been published by a Royal Commission in 1712. Cause and effect are very confused in the quarrels and disputes that led from Flamsteed's being eager and willing to publish his results in 1703 to his bitter resentment of all who were involved, especially Halley and Newton, by 1712. Flamsteed declared that the 1712 edition was so inaccurate and incomplete that he recalled as many copies of the book as he could find, in order to destroy them. He then spent the rest of his life, until his death in 1719, preparing his own version as he wanted it to be.

When Flamsteed died, his assistant at the Observatory was Joseph Crosthwait, a Northumberland boy who had been trained on the job and was now a competent astronomer. Crosthwait, too, was sufficiently loyal to his former teacher and employer to devote several more years to supervising the printing of the book, even turning down the offer of another job in order to finish it.

Crosthwait's correspondence shows that there were many more troubles and annoyances to be met in carrying out Flamsteed's wishes. In his will, Flamsteed had left to his wife £120 per annum as well as the £50 per

ATLAS

COELESTIS.

By the late Reverend

Mr. *JOHN FLAMSTEED,*

REGIUS PROFESSOR of ASTRONOMY at *Greenwich.*

LONDON, Printed in the Year M.DCC.XXIX.

Figure 4.2 The title page of Flamsteed's *Atlas of the Heavens* finally published ten years after his death. Royal Greenwich Observatory.

annum that she had brought him. All his books and instruments were to be shared between her and his niece, Katherine Herring, who had married the former assistant James Hodgson, later master of Christ's Hospital. Mrs Flamsteed and Mrs Hodgson were joint executrixes and against them the Board of Ordnance brought a bill to repossess the instruments that Flamsteed had bought with his own money. With Crosthwait's help the attempt was repulsed and they struggled on with the printing until 1725.

There was not only the printing to manage. In 1725 there were still in circulation a few copies of the 1712 edition of the *Historia Coelestis* that Flamsteed had wished to destroy. A letter survives from Margaret to Dr Mather, Vice Chancellor of Oxford, asking him to remove one remaining copy of the 1712 edition from the library.

No one would deny that Margaret Flamsteed gave loyal help and service to her husband while he lived and after his death. Unfortunately, her image is a little tarnished at the end. When she died, she left nothing to Joseph Crosthwait, who had done most of the work involved in supervising the printing. Flamsteed had left him nothing in his will and he had been paid nothing. He wrote bitterly to Abraham Sharp: 'For all my time spent and all my own expenses in attending the printing and maps I never had any allowance besides losing two places that were offered me; one in the Ordnance Office at £80 per annum which I refused at her request. What has induced her to act so dishonestly by us at last except it was that she had no further occasion I cannot apprehend' (29 August 1730).

James Hodgson, who did benefit from Flamsteed's will, is the only one of Flamsteed's assistants whom we know to have been married. Although his wife was Flamsteed's niece, Anna, she seems not to have been involved in his work. Flamsteed never mentions her in a work context. Later, as the wife of a headmaster, she could probably not altogether avoid involvement. The other assistants, Denton, Smith, Sharp, Stafford, Leigh, Weston, Ryley, Witty, Woolferman and Crosthwait, were all unmarried while they worked at the Observatory, as far as we know.

Among the correspondents, we can be fairly confident that Stephen Gray never married. In his youth, he never made any mention of a wife while he was working as a dyer. This, in itself, would not be very strong evidence: he was writing impersonal letters, and personal details are included only if they impinge on scientific activities. A wife might be completely uninvolved, although the most domestic of wives could have an effect on her husband's activities. Oughtred's wife refused to allow him candles to work by after supper, and in that way made her contribution to what was done—or rather, not done—in science.

Gray was certainly not married by the time he asked to be admitted to the Charterhouse in London. In 1711, he wrote to Sloane to say that he was finding the work of dyeing too much for him. He had back trouble, and more importantly, one suspects, not enough time for astronomy. As a way out, he suggested that someone influential, like Sloane, might be able to get him into the Charterhouse, a home for penurious old men. The constitution states clearly: 'in order to be admitted a Pensioner at this Hospital he should at the same time be obliged to declare upon oath that he is not a married man.' The Charterhouse specifically made provision for respectable old men only: 'It is constituted and ordayned by the Consent of all the said Governors that there shall noe rogues or Common Beggars be placed in the said Hospitall but such poore persons as can bring good testimonye and certificatt of their good behavioure and soundness in Religion and such as have been servauntes of the Kyng's Ma' either decrepit or old Captynes either at Sea or Land, Souldiers maymed or ympotent decayed Merchaunts men fallen into decaye through shipwrecke casualtie, Fyer or such evill Accident.' The inhabitants were likely to be men with a past. In Stephen Gray's case there was still a considerable future, as he settled down to do some of his most useful scientific work after his admission.

The history of the Charterhouse allows us a closer look at the domestic arrangements for the Pensioners than is available for the other astronomers, whether married or not. The Governor's Order Books show the day-to-day running of the establishment, and what was acceptable and what was not. The original charter of 1614 provided an allowance of £5 to each brother per year. In addition to this, he had a room opening off a staircase round a quadrangle, very much like an Oxford college. At midday, dinner was provided, consisting of only meat, bread and ale. Eventually, in 1737, a year after Gray's death, the governors ordered 'that ye officers together with Dr Hall ye Physician do consider of the Diet of the Pens^ers in respect to Salleds Greens and roots and to make such alterations forthwith as they shall think fitting'.

Supper was provided at six o'clock, but there was no breakfast. An order issued in 1730, while Gray was still there, indicates both the hard living conditions and the monastic discipline exercised by the Governor: 'they have no allowance for breakfast, it is a great hardship to fast from supper at six o'clock to dinner . . . and therefore humbly praying us to make them some allowance for breakfast. We do order that ye said Pensioners if they think fit may take the broth provided for them at dinner in ye hall and carry it to their chambers for breakfast the next day.' The brothers had to contend not only with a meagre diet and disciplinary restrictions but also with the depradations of the poorly paid staff. A custom grew up that bread should be cut up in the kitchen before it was served in hall. At this point, the kitchen staff removed a proportion of it as their own right. In 1714, the pensioners filed petitions asking that the bread should not be cut up before it was served. The petitions were granted. The kitchen staff objected vehemently and launched counter petitions to restore their privilege of 'chipping the bread'. They were denied, and, on this occasion, the brothers triumphed, being allowed to eat their full portions.

For the seventeenth century this seemed adequate diet, and the brothers probably lived in comfort, compared with the growing number of paupers in the country who lacked adequate food and housing or the hope of ever improving. Those who could afford it ate large quantities of meat, and more exotic eating and drinking habits began to creep in during the eighteenth century. Tea, for example, halved in price, and sugar grew continually cheaper. Coffee houses provided coffee, chocolate and tea. Hooke, we know from his diaries, had a great taste for chocolate. By the 1750s these drinks had spread to all classes in society. 'Even a common washer woman thinks she has not had a proper Breakfast without Tea and hot buttered white bread' (Porter 1982). Hooke records very carefully which food and drinks made him ill and which seemed satisfactory. From his records of the 1680s he appears to have eaten a varied diet in which surprising things agreed with him, for example, 'Eat nuts and brandy. Agreed well and slept.'

As a place in which to practise science, the Charterhouse presented various problems. Stephen Gray, in his later years, ceased to work on astronomy and concentrated instead on experiments with electricity. He seems to have circumvented rules in order to string up silk threads in the quadrangle through which to transmit electric charges. In fact, Gray himself talks of parties of ladies and gentlemen coming on outings to visit him and watch his astonishing experiments: 'I have now and then some Companys of Gentlemen and Ladys come to me to be entertained with my Electrical Experiments I am in hopes when the days are longer and the weather better I may have more than I have at present' (14 February 1736). As this was written only nine days before he died, Gray was obviously very active, and must have enjoyed his fame or notoriety.

Astronomy was an increasingly fashionable subject in the eighteenth century. Cotes' difficulties with crowds of visitors trying to watch his observing have already been noted. Flamsteed had often been bothered by visitors of various sorts coming down to Greenwich to see the Observatory. He was troubled by casual visitors and groups of ladies and gentlemen with nothing more than idle curiosity and no specialist knowledge. He was rather proud, however, of some visitors. For example, the Tsar of Russia, Peter the Great, visited London in 1698. He went down to Greenwich in February and Flamsteed entered a note in his observing book: 'Serenissimus Petrus Muscoviae Zarus observatorium primum venit, Lustratisque instrumentis abiit [Tsar Peter of Moscow came to the Observatory for the first time and after looking at the instruments he left].'

The Tsar must have found his first visit more interesting than the note implies, because he came back in March and watched Flamsteed observing, as another note in the observing book testifies. In all, he came four times.

Visitors seem not to have disturbed Gray. In fact he liked them. One rule that might have interfered with his work was the restriction on travel that required pensioners to have permission in order to stay away from the Charterhouse for more than a short period of holiday. We know that Gray spent time 'in the country' with his friends John Godfrey and Granville Wheler in Kent. Permission must have been reasonably easy to obtain. Living gowns, like those worn at the universities, were required, and must have contributed to an institutional atmosphere. Yet there was still privacy and the opportunity for companionship. After Gray's death, a fellow pensioner, Thomas Childe, wrote to Sloane to ask for money in virtue of having known Gray. In spite of the mercenary purpose of the letter, the picture of companionship that it gives is attractive: 'Mr Gray would smoke two or three pipes and gave me a great deal of delight and satisfaction in his very agreeable conversation.' The two men 'did eat and drink together on Saturdays and Tuesdays'.

Gray, at least, ended his life with some companionship, although he seems to have been a lonely, rather shy man in his youth. Abraham Sharp, on the other hand, grew more lonely as he grew older. In his youth, he travelled from Yorkshire south to London to work for Flamsteed. Later he spent some time working as a nautical instrument maker at Portsmouth. Life must have been full of interest and he must have met many people. Then, in 1693, his brother the Reverend Thomas Sharp died. Abraham returned home to Yorkshire to look after his sister-in-law and his young nephew, who was studying medicine. Within months the nephew also died, leaving Sharp without 'the only person here with whom I could have any agreeable converse'. He had no-one left to help him to make observations or to count the clock, but the loss of conversation affected him most. He was more than ever anxious to continue his correspondence with Flamsteed: 'I desire to hear often from you but am loth to put you to ye expense of my answers to you when my letters shall bring you nothing is worth the postage.' He suffers from a sense of his own unworthiness but clings to his one link with the world of action: 'I would gladly still if possible retain your correspondence it being what has ever afforded me as great satisfaction as any earthly thing for cultivating whereof I have little to offer' (8 December 1702).

Sharp embraced loneliness and made no compromises with it. In a room on the first floor of Horton Hall was his study with his oak desk and shelves of books. So many hours were spent at the desk that the servants reported little hollow cavities in its surface, worn there by the pressure of his elbows.

These many hours were spent alone. No-one was admitted, even to bring him food. He had a little serving hatch made in the wall and, like an enclosed monk, he worked on undisturbed while food accumulated in the wall. He complained to Flamsteed that there was no-one with whom to discuss mathematics and astronomy. The only two whose conversation gave him any pleasure were a mathematician called Dawson and an apothecary called Swain. A somewhat bizarre state of mind can be seen in the system that Sharp evolved to make sure that he was not interrupted by harsh sounds when he wanted to concentrate. If either of these two came to talk to him, they were required to stand outside and rub a stone against the outside of the door. That Sharp grew as lonely as Newton had been at Cambridge is hardly surprising. Intellectual equals were impossible to find and neither man wanted to spend time on ordinary human companionship.

Sharp never married but he showed fatherly interest in his niece's daughter, Faith. She was his only surviving relative and, like Hooke for his niece and Newton for his half-niece, Catherine, he seems to have felt affection and concern for her. Letters written to her while she was a schoolgirl at York enquire anxiously for her health, but show a tendency to reprimand her for faults in the replies, such as shaky handwriting. Sharp helped to negotiate her marriage and continued to correspond with her and her husband.

The astronomers seem to have been mostly solitary men. With the notable exception of Halley, they either did not look for, or were incapable of finding, a large number of friends and casual acquaintances. Some of the time they were content with their work and a few contacts with others who understood what they were doing. Some of the time they felt isolated, aggrieved, unappreciated. Yet it would not be true to say that any one of them was incapable of forming relationships. Each astronomer had some closeness to another person, though probably less than would be found in a random sample of the population. An interest in astronomy might therefore have taken up the energy that others could put into human relationships.

5

Money matters

The traditional sources of money for intellectual activities were the Church and the land. Among the astronomers in the time of Newton, few gained support from family property. John Wallis' mother was descended from a land-owner. Brook Taylor's father owned a reasonable estate in Kent, and both were able to be educated and live as academics from their father's property. The same applies to William Molyneux, whose father, Thomas Molyneux, was an Irish land-owner with property in several counties and a town house in Dublin. Thomas Molyneux's estate enabled his son to go to town and mix with the urban population of Dublin, and to go to school and university. Lastly, Richard Towneley inherited a family estate in Lancashire and was able to live from its proceeds.

But in the eighteenth century the sources of money were changing. The population was beginning to expand and trade was the way to go—Defoe wrote that 'an estate is but a pond but trade is a spring' (Porter 1982). London was big, and throughout the eighteenth century it grew bigger and richer. In 1700 London used 800 000 tons of coal. By 1750 the amount had doubled to 1 500 000 tons. A small part of its wealth and expansion was formed by the instrument makers and the booksellers who served the astronomers.

What of the astronomers themselves? Their share of wealth was rarely very large and, when it came, it was rarely from personal involvement in trade. However, some astronomers, like Flamsteed and Halley, owed the money for their education or leisure to their fathers' successful trade, and Abraham Sharp's father had made his money in the clothing trade. Stephen Gray's father was a dyer but obviously had not been sufficiently successful to allow his son independence. Both Sharp and Gray were expected to go back into their fathers' trades, but Sharp, from the more prosperous northern clothing industry, was able to break away.

Among earlier seventeenth-century astronomers, Isaac Barrow's father was a first-generation tradesman. His grandfather, Thomas, was a land-owner and Justice of the Peace in Cambridgeshire. He never intended his son to enter trade but, according to Aubrey, the two Barrows, both called Thomas, could not bear each other's company, and so Isaac's father went to London and was apprenticed to a linen draper. In the Civil War his aristocratic blood sent him off to fight for the King, and Isaac, who was at Oxford, was left to fend for himself. When the wars were over Thomas Barrow promised to try to find his son £20 a year so that he could stay at Trinity College. This was a large sum for a linen draper whose annual income might be as little as £40. Joseph Massie estimated the average incomes of families in different jobs and social groups in the 1760s. 'Lower merchants' by this time are listed as earning possibly £200 and upwards. A hundred years earlier, incomes were lower but prices, on the whole, did not increase greatly. Taking the value in 1700 as 100, the index for consumer goods was actually lower in 1750 at 92.

A large number of astronomers had fathers who were ministers of the Church but, of these, many died while their sons were still young. The result is that very few astronomers embarked on the pursuit of astronomy with any financial backing at all. To begin with, they all had to eat, and they present a variety of solutions to the problem of finding food, shelter and the necessities of their usually rather austere lives.

Nearly all the astronomers, including amateurs, had a paid job of some kind from which they derived a living of sorts. The cases of John Pell and Michael Dary who, Aubrey tells us, died in poverty are an extreme, but very few seem to have lived in much luxury. The jobs themselves vary considerably, although a living in the Church was a favourite, particularly in the seventeenth century, as the hours were highly conducive to the practice of astronomy. Derham, Flamsteed, Pell, Pound, Thornton, Ward and Wilkins all had livings, and Ward and Wilkins both became bishops.

Since the Church of England was thus involved to some extent in subsidising astronomy, it is tempting to ask whether the Church was getting its money's worth from its servants. We have detailed information about Flamsteed's activities as rector of Burstow, and some information about the amount of time that he devoted to this job. An examination of his correspondence makes it fairly clear where he was prepared to put most of his effort, although he obviously cared about his responsibility in the Church.

Flamsteed took his degree of MA at Cambridge, in 1674. In his auto-biographical notes he records that he took holy orders in 1675 at Ely House from Bishop Gunning—who was very anxious to talk about the new philosophy, although he, personally, maintained the old. Flamsteed had to insist on taking orders against the advice and wishes of his family and friends. The implication is that he was eager to enter the Church for its own and his own sake, and should not be accused of hypocrisy.

Between taking orders and finding a Church living, however, the providence of God and the good offices of Jonas Moore intervened. Flamsteed was offered and accepted the post of Astronomer Royal, and was installed at the new observatory in August 1675. Existing on the small salary of £100 per annum, out of which he had to pay for instruments and assistants, became Flamsteed's greatest problem. His aim and the King's wish was that he should produce a new, accurate catalogue of the stars, but the practical difficulties of financing instruments and assistants prevented him from getting on with his work as he wished. Flamsteed's correspondence includes letters to civil servants, in which he tried to persuade them to pay his salary.

In 1678, an opportunity to escape from these pressures presented itself. Dr Bernard wrote from Oxford to encourage Flamsteed to apply for the post of Savilian Professor of Astronomy, which had just become vacant. Flamsteed declined to apply, but said that he was very far from happy with his job at the moment. Working conditions were poor: 'I am as weary of the place as you of yours', and he would prefer a healthier environment than Greenwich with 'lesser or fewer distempers'. He would move, however, only to something which would be 'more useful in the world' and would promote 'more glory to my maker'.

Flamsteed might, if elected, have enjoyed the peace of an academic chair in which teaching and lecturing could be kept to a minimum. Isaac Newton held a professorship but appears to have avoided teaching almost entirely. The temptation of such a life, however, was not strong enough for Flamsteed. His resistance may also be a comment on the desirability to a practising astronomer of working at one of the universities. A theoretician like Newton might have been satisfied, although even Newton seized the offer of a job in London when it came his way. This reluctance is thrown into relief by the next offer of a job that Flamsteed received. In 1684, the Church living of Burstow became vacant and Lord North, in whose gift it lay, offered it to Flamsteed, who accepted it.

Burstow was a small parish in the Weald on the borders of Sussex and Surrey. In 1700 it consisted of 606 persons. The income from the tithes and living was very small, about £150 a year. Moreover as Flamsteed did not stay in residence to take services and perform parish duties himself, he had to pay a curate about £40 to do the work for him. He also had to maintain buildings and work the lands. As a result the income derived from the living was mostly in the form of goods and services. Since, at Greenwich, he had to feed his own extended family, his servants, and whatever boys or young men he was teaching, supplies of food must have been a valuable form of income. Collecting the tithes was not easy. One parishioner tried to take advantage of the absence of the rector to persuade the curate, then Timothy Stileman, that Flamsteed had agreed to remit his tithe, and therefore he would no longer pay it. A sharp letter from Flamsteed at the Observatory to Stileman made very clear that the tithes were to be paid in money or wood. If the offender did not pay, he would be taken to court. Flamsteed, present or absent, was not to be cheated.

Flamsteed went to Burstow about twice a year, on average. Each time, he stayed for a few weeks to take services and see how things were going. His letters to the curates show that he took a very detailed interest in the care of the Church lands. He sent instructions on ploughing, for example, and gave directions for the cutting of the copse and for mending fences to keep the cows from breaking in and trampling the young trees.

At harvest time, Flamsteed liked to make a personal visit. In 1705, for instance, he wrote to Newton: 'I must go into Surrey to reap my harvest as I usually do about this time.' This he meant literally: he was not thinking of souls.

A personal visit was all the more desirable since the curate, Mr Stileman, was not giving satisfactory service, and Flamsteed had to keep a close eye on him. His letters often sound irritated, and even take on a stern note of rebuke. Mr Stileman liked to be elsewhere than in Church on Sundays. Flamsteed was displeased to find that the service had been neglected: 'I am sorry you engaged yourself to serve Mr Jones the first Sunday afternoon that I was absent. Pray do not absent from my people again without some very extraordinary occasion any Sunday afternoon or morning' (22 November 1706). This instruction was ignored, and Flamsteed had to write again with fiercer rebukes about services not taken and parish duties neglected.

Mr Stileman gave other causes for complaint. Apparently, in March 1706, he wrote to Flamsteed that he had no money and was waiting to be paid. Flamsteed rebuked him for having already spent the money he was given, in addition to his salary for the Christmas quarter. Here Flamsteed demonstrated some management skill. He appealed to the curate by praising him: 'I cannot think so good a manager as you are quite broke', but ended with the disapproving comment 'I fear you are given to feast'.

The picture of Mr Stileman's failings as a curate is completed by his failure to travel to Greenwich to report to his rector as often as was required. Certainly, the weather made country roads very bad at times. Burstow is more than thirty miles from Greenwich, and for much of the way there was nothing better than a bridle path or track. In November 1706, Mr Stileman said that he could not get to Greenwich, which provoked the reply: 'Mr Green was lately in town if the ways are not so abhominable bad but he can travell them I suppose they will not be worse to you then to him and you are not so heavy as he is this is no good excuse.'

Flamsteed himself found the travelling difficult and, in spite of his arguments to his curate, found that he could not make the journey himself in bad weather. In March 1706 he had been compelled to write that he could not travel to Burstow: 'By reason of the badness of the ways I dare not yet take my journey to Burstow for I am scarce yet well of my lameness and ye paines of my feet and I am much afraid the cold and wett I must necessarily meet with below the Hills may cause a Relapse.'

Flamsteed was a man of sixty and could reasonably expect his younger curate to come to him. The parishioners, therefore, must have seen less of their rector than Flamsteed would probably have wished. But, present or absent, he was concerned for their welfare. He added a postscript to the letter of March 1706: 'I think it be my turn to choose the Churchwarden, Pray look into the book and see and if it be let me know by the Reigate post, and who of my Neighbours hath been longest out and I will order you who to pitch upon. And if you want directions in any other affairs inform me.' Nevertheless, there is no denying that he was an absentee.

The practice of appointing absentee vicars was common in the eighteenth century. However, James Bradley, made Astronomer Royal in 1742, was offered the living of Greenwich to supplement his income and, surprisingly, turned it down on ethical grounds, because he did not wish to be an absentee. The parish of Burstow was therefore no worse off than many other late seventeenth/early eighteenth century parishes where latitudinarian vicars and rectors preached 'be not over zealous' while their congregations slumbered happily in front of them. Parish business continued, and, although tithes were strictly exacted, some money returned to the parish in wages for ploughing and working the Church lands or glebe.

Mr Stileman was not a conscientious curate, and it is difficult to imagine his visiting the sick or relieving the poor and needy. His successors were not without their problems; curates were as much of a problem to the clergy as servants to the aristocracy. In 1715, Flamsteed dismissed a curate whose principles were too High Church for his own austere tastes. He found it difficult to appoint a replacement who would not be 'tainted with the same principles'. Bishop Kennet wrote to Flamsteed as though the task of finding a satisfactory curate were very arduous: 'I have been so anxious to provide a safe and honest curate for you and I think at last I have fixed upon a man against whom I know of no objection' (1 November 1751).

Other astronomers who were clergymen also seem to have devoted small proportions of their energies to their clerical duties. The one who made the most successful career in the Church was probably William Derham. He was given the living at Wargrave, in Berkshire, in 1682. He progressed in 1689 to the more valuable living at Upminster in Essex. He gained by this move both a better income and proximity to London, which enabled him to make himself known to the Royal Society. He involved himself to the greatest extent after 1702 when he was elected Fellow. He wrote on many aspects of science in connection with theology and was generally respected for his knowledge and his personal merits.

A closer look at comments on his work as rector of Upminster shows a less positive image. He spent a great deal of time in reading and writing, and was undoubtedly very learned in theology, as well as natural science, but one biographer wrote that 'it sometimes happens that clergymen of the greatest wisdom, learning and merit are far from being good preachers. Dr Derham is understood to have made but a very poor figure in this respect, and to his other defects in the pulpit was added some disadvantage with regard to his person for he was very long necked' (Chambers 1820). The picture is of a very awkward figure preaching academic and philosophical sermons over the heads of his parishioners.

In another way, Derham did meet their needs. He was one of the few astronomers in this group who was able and willing to work as a physician. He may have been more helpful to his parishioners' bodies than to their souls, and his reputation was certainly higher for his medical cures than his pastoral care.

Derham's inept preaching was no obstacle to promotion. He was known and liked in court circles, and when George I ascended the throne in 1713 Derham was given a plum: the chaplaincy to the Prince of Wales (afterwards George II). As there was a serious rift between the King and the Prince over the upbringing of the Prince's children, Derham may have found the post less easy to manage than he had hoped. He seems to have kept out of trouble because, in 1716, he acquired the additional honour of a canon's stall at Windsor. When George I died unexpectedly in 1723 and the Prince of Wales became George II, Derham was still in office (and competing with the vicar of Bray in staying power!). He survived in material comfort until his death. An interest in medicine and science was certainly no encumbrance in moving up the Church hierarchy.

James Pound (or Pounds) was another who managed to combine astronomical work with clerical duties (both in dangerous circumstances), although he never achieved as high a rank as Derham. Pound's first position was certainly more challenging than most. He had taken a medical diploma as well as a degree at Oxford, and with these qualifications he was appointed a chaplain for the East India Company. In 1699, he set sail for Madras and from there went on to a settlement on the river Cambodia. He intended to make use of his time there for astronomical observations, but problems beset his projects. In January 1700, Flamsteed wrote to him about a quadrant that he was sending out for Pound's use. Flamsteed himself was taking the trouble to buy and test the quadrant, and was encountering the usual difficulty with ignorant instrument makers. He was also sending some calculated places of stars to give Pound the foundation from which to make observations of positions of stars in the southern hemisphere below about declination $-30°$, which was as far as Flamsteed could reach from Greenwich.

Halley, at St Helena, had begun this task of cataloguing southern stars, but had not gone very far because of cloudy weather and the inconvenient site of his instrument, which was a considerable distance away from his house. Moreover, Flamsteed added, Pound was much more likely to be successful because he was esteemed 'a very good man'. Halley, on the other hand, was 'blasted by reason of his impiety and profaneness and that he sought his own reputation more than anything else'. Here Flamsteed gives vent to his hostility to Halley while trying to secure the loyalty of Pound. The important sentence comes at the end: 'I expect you should send me ye measured distances as you take them for haveing contrived and formed the instrument I have a right to them.' In case this sounds a little blunt, he adds that they will then of course be available for the benefit of all seamen and, in fact, for all nations who navigate the southern seas.

As chaplain for the East India Company, Pound ran into difficulties. In 1705, the natives at Palo Condore mutinied. Pound was awakened in the early morning of 3 March by the noise and smoke of a fire and by the screams of his companions being slaughtered by mutinous natives. Pound, with a few others, did not linger. As they discovered later, the governor had been shot and stabbed with a poisoned arrow. They could see and hear the whole fort burning around them as they made their way to a boat. Wearing only his nightshirt, Pound was one of the few who succeeded in escaping and sailing to Batavia. Escaping with such speed allowed no time to grab breeches, money, papers—or even astronomical observations! All the results that he had achieved up to that moment were left behind to burn.

In 1705, on 7 July, Pound wrote to Flamsteed from Batavia, which he had reached in safety, to say that he had lost everything and must return to London, 'having neither instruments nor Money nor Books'. If his instruments had survived, he would have stayed for one year, 'at my own expense', to complete the observations for a catalogue of southern stars, but without instruments he could do nothing. Since the quadrant had taken three years to reach him, we can understand his unwillingness to wait for another one to come, even if he could have afforded it. Considering all the problems, he told Flamsteed that he was returning to Condore in the ship called the *Caesar* to see whether there were any English survivors there or any goods to be recovered.

Pound was not able to resolve the problem of the loss of money and instruments and was compelled, as he had told Flamsteed, to return to England in 1706. The following year he began his upward progress in the Church. He had already, in 1699, been elected a member of the Royal

Society, and had made himself known to men in whose hands rested the power of patronage. In 1707, Sir Richard Child presented to him the rectory of Wanstead in Essex. Wanstead provided him with enough money to marry and to begin to acquire instruments for observing. In 1710, he married a widow, although she had brought no fortune. She bore one son, then died in 1716. After her death, Pound's material position improved greatly, as opportunities for promotion appeared. The Church provided opportunities for advancement. The living of Burstow became vacant on Flamsteed's death in 1719. Pound was by this time well known and respected. The intervention of Lord Chancellor Parker on his behalf acquired for Pound the offer of the living. Having no scruples over absenteeism or pluralism, Pound accepted it gladly and thus greatly expanded his income. He already had a good selection of instruments and would be able to buy more. Halley had said in 1715 that Pound was 'furnished with very curious instruments'. In 1722 the likelihood of financial restraints was completely removed by a very judicious marriage to the daughter of a local land-owner. Elizabeth brought with her a fortune of £10 000. Her brother had made a great deal of money by investing in South Sea stock. By this marriage Pound joined the few astronomers of the period who had plenty of money and no problems over acquiring instruments or hiring assistants, other than finding suitable ones.

As the income from a single parish was likely to be low, some clerical astronomers tried to supplement their incomes by private enterprise. The Reverend Stephen Thornton ran a school in his parish of Ludsdown, in Kent. Anxious to join the astronomical community, he wrote a few letters to Flamsteed, in one of which he revealed the insecurity of trying to make money by running a boarding school: 'We have been in some hurry as one of my young Gentlemen has ye small pox and has made some others fly and we have been put to it to secure the rest.'

A small country parish where the income was low could provide barely enough for a living for a man and his family. Presumably the poorest priests are not represented among Flamsteed's correspondents. The Reverend Matthew Wright, living in Cheshire, who describes himself as 'an unworthy priest of the Church of England', had somehow acquired a 16-foot telescope. He wrote to Flamsteed like many others asking for forecast positions of Jupiter's satellites so that he could make some useful observations.

On the whole, the astronomers connected with the Church of England were well able to provide for themselves and their families in those cases where there was a family. In some cases, a university job was combined with holy orders. Seth Ward was Savilian Professor at Oxford and bishop of Sarum. John Wilkins was appointed Savilian Professor of Astronomy at Oxford on Ward's retirement. Like Ward, he found a university job difficult to keep through the changes of the Civil War and Restoration. Because of his loyalty to the Commonwealth he lost the Mastership of Trinity College, Cambridge, at the Restoration, despite petitions from the Fellows that he should be allowed to stay. After this he made his way up in the Church to become bishop of Chester in 1668. He continued to be a highly respected intellectual (he was one of the first secretaries of the Royal Society), but his main work in astronomy was carried out while he was in the university. During his period in the Church, he became more interested in the abstract questions of philosophy than in experimental science, and he published a dictionary of philosophical language.

John Wallis, by contrast, was a university man all his life. He was appointed Savilian Professor of Geometry in 1649 by Cromwell, even though he had signed the Remonstrance against the execution of the King. As a mathematician, his interest in astronomy was peripheral; one major *coup* was a practical joke, by which he invented a cipher capable of so many different interpretations that he was able to claim priority for Huygens' discovery of Titan, largest of the moons of Saturn. Huygens accepted Wallis' claim with a good grace but was somewhat irritated to find that the whole thing was a joke. Wallis never climbed on the promotion ladder provided by the Church but found that his duties as Savilian Professor allowed him to live the kind of life that he enjoyed.

In addition to his income from the professorship, Wallis managed to acquire the extra income from being Keeper of the University Archives. In this capacity he fell foul of Anthony à Wood, the antiquarian, who had been accustomed to having the free run of the archives. Wallis took the key away from Wood, accusing him of tearing the title page out of a book. Wood denied it and appealed. Wallis agreed to let him in again if the Vice Chancellor would consent. Wood went to see the Vice Chancellor who said he would have to see Wallis first. At this point, Wood gave up, but he carried his antipathy for Wallis with him for some time. He accused Wallis of keeping his professorship by being loyal 'to Oliver, to Richard, To Charles II, to King James and to King William'. He accused Wallis of obtaining the archives job by breaking the University Statutes. Aubrey, too, takes this view, saying that the Savilian Professor is not allowed to take another job.

In spite of the protests, Wallis was confirmed in both his posts by Charles II, partly, no doubt, because he had shown the courage to sign the Remonstrance against the execution of Charles I and so had shown some preference for the Royalist side. He was also used at various times as a cryptographer because of his known skill in deciphering. This kind of work was not usually well paid and was sometimes not paid at all. When he was employed by the Earl of Nottingham to work on behalf of King William, he had to ask for 'some better recompense than a few good words, for really my Lord it is a hard service requiring much labour as well as skill'.

University jobs provided an income increasingly, and no doubt partly for this reason men who were to make a name in astronomy were likely to spend at least part of their working lives in a university. Flamsteed, as the first professional astronomer devoting his whole career to observational astronomy, was also exceptional in refusing to consider university work. The two Gregorys, Wallis, Newton, Cotes and Keill are representative of professional academics who spent at least part of their careers in a university chair and did not hold a living in the Church.

Working in a university was a largely undefined activity in the seventeenth century, as in the twentieth. A man who had sufficient prestige to be of benefit to the university by his mere presence could manage to avoid all but a little teaching or lecturing. In fact, after the Restoration, teaching was at a low level in both the universities (Westfall 1981). At Cambridge, divinity acts and exercises were required of all Masters of Arts of more than two years' standing, but increasingly often they were neglected. Westfall points out that Newton completely ignored these statutes while he was Lucasian Professor. Advancement up the steps of college hierarchies depended entirely on seniority. A fellow, once appointed, could sit out his time until he had served sufficiently long to acquire a college living. At this point he could marry and have the benefits of family life, if that was what he wanted. In either case there was little demand made on him for academic performance. Dr Johnson, in 1754, pointed out that a man who had been a student at Pembroke College with him had stayed at Oxford to 'feed on a fellowship' but such a life had not improved his literary performance or anything else about him (Boswell 1965).

For Newton, the life of a don was ideal. He had money to buy the equipment needed for the chemical experiments on which he devoted much of his time and few other demands were made on him. In fact, the isolation in which his colleagues left him was probably an advantage for his work, as the other fellows were not likely to contribute much to his thought or stimulate new enquiries. Outside correspondents like Hooke and Halley were chiefly responsible for doing that. The fellows of his college were able to help Newton best by not treading on the drawings that he made with his stick in the gravel of the walks in the college garden.

Convenient though Cambridge undoubtedly was for Newton, it provided in every way a sanctuary rather than a stimulus. When an opportunity came to leave for London he had no hesitation in going. A position was found for him as Warden of the Mint. The salary was good and the job could be regarded as a sinecure. London was the place where lay interest and excitement for a scientist. Newton rejected the comfortable college living in the Cambridge countryside to which his seniority entitled him and, instead, left Cambridge with scarcely a backward glance.

Newton, Halley and William Molyneux all held posts in the Mint: Newton was in London, Molyneux and Halley at Chester. The precedent among Wardens of the Mint was that the job should be a sinecure. Previous Wardens had regarded the £400 salary as an honorarium. Newton chose, instead, to supervise in person the complexities of the recoinage that was already under way when he arrived. He devoted a considerable part of his time and energy to calling in the old coins. In 1696 he used his position to appoint his younger friend and colleague, Edmond Halley, to the post of Deputy Comptroller of the Mint at Chester, one of the five temporary Mints set up in the Provinces to help with recoinage. William Molyneux was Comptroller. Halley ran into various difficulties on this job because he, too, attempted to carry out the duties of the position and not use it as a mere sinecure.

The life of an official of the Mint could involve some danger. The Warden at Chester discovered some irregularities in the distribution and collection of money. As a result he was challenged to a duel. Fortunately for him, the challenger lost his nerve: he arrived early, thus preserving his own honour, and left early, before it was time for the duel, thus also preserving both lives.

Halley ran into some trouble of his own. He was accused of mixing schissel alloy with the metal of the coins. He appealed to Newton for assistance. His intention was to resign from his post but at the same time to prosecute his accuser and the Warden's assailant so that his leaving would not be interpreted as flight. Newton's response was to try to find Halley a new job. He had already offered him a post as some kind of engineer. When he received Halley's letter, Newton was able to offer him a job teaching mathematics and engineering to the army for 10 shillings a day. Halley turned down both these offers and sat out his time at Chester until the Mint there was closed down a year later and he was able to return to London with dignity.

On the fringes of academic life were the jobs provided by professorships at Gresham College, founded for the education of adults in London. A Gresham Professor had none of the status conferred by a university community. He had the disadvantage, compared with his Oxford and Cambridge colleagues, that his main function was actually to deliver lectures. Nevertheless, many who held the posts were not conscientious. When Flamsteed agreed to deliver astronomy lectures for Professor Walter Pope, who was too ill in 1681 to do so himself, he made some comments that point to a sad state of decay of Sir Thomas Gresham's original intention: 'Let the others break faith or excuse themselves or evade their duty. As for myself if I only fulfill no more noble task than obeying the laws of the place and your injunctions I shall consider myself never to have done anything more distinguished or praiseworthy.'

The casual attitude held by many of the professors to delivering the lectures is more easily understood in the light of other comments made on the attendance at the lectures. For example, at his first lecture Flamsteed had 'but a slender auditory', and in some cases the manuscript has the comment 'No body came not read' on the bottom. Flamsteed's sense of persecution appears in one case when he listed three attenders 'with the Blewcoats' (boys from Christ's Hospital) and adds 'Mr. Hookes' designe'.

If all the lecturers suffered from such poor attendance, their lack of enthusiasm is understandable, but in such cases the difficulty lies in knowing which came first: poor audiences or reluctant lecturers.

Hooke was appointed Gresham Professor of Geometry in 1665. He already had apartments at Gresham College, which had been assigned to him when he began his job as curator of experiments for the Royal Society in 1662. The Royal Society job had been at first temporary, but was made perpetual with a salary of £70 per annum. Hooke had continuous trouble over the salaries from all his jobs. The Royal Society paid him grudgingly and often in arrears, mostly because they depended on subscriptions and genuinely did not have much money. When Halley was Secretary in 1687, the journal of the Royal Society records: 'Resolved Halley be given 50 copies of *History of Fishes* instead of £50 salary.' Halley was expected to make the best bargain he could with the bookseller to dispose of the books which were obviously not selling well. Hooke was often paid in cash when he was paid, but the event was sufficiently remarkable to be recorded in his diary. Hooke's election as Gresham Professor of Geometry was disputed. In fact, Arthur Dacres was declared to be the legally elected professor in 1664. Sir John Cutler, on hearing this, offered Hooke a salary of £50 per year to deliver lectures. Hooke accepted this offer but disputed the Gresham decision. The election of Dacres was declared invalid and Hooke was installed as Gresham Professor in his own right.

Unfortunately, Sir John Cutler proved as unreliable as the subscribers of the Royal Society. He showed no great determination to keep Hooke for the rest of his life, although Hooke faithfully delivered the sixteen Cutlerian lectures every year as he had promised. In 1668 he asked for and received a statement from the Royal Society to certify that he had delivered the lectures. He took the certificate with him to the Court of Chancery to sue the heirs of Sir John Cutler for the salary owed to him. The case was opened in 1688, but was decided in Hooke's favour only in 1696, when he was eventually paid.

Hooke had various sources of income during his life and each was a problem to him. After the Great Fire, he was appointed one of the surveyors to work with Wren on surveying and rebuilding the City of London. For this work he was to be paid £150 per year by the City Council. This salary was often in arrears. In 1675 he recorded: 'May 25 Sir Chr. Wren unwilling to let me have any money. He seemed jealous of me. Jun. 1. At Sir Chr. Wren but noe money nor favour.' Persistence, however, brought its reward. Hooke continued to press for his salary and, on 26 June, after waiting a month, he was able to record: 'To Sir Chr. Wrens stayed till 3 to see payment of money. Received for my salary £50.' But every time Hooke had to struggle for his salary. At the end of 1676 Wren owed him £150.

Others paid Hooke for his surveying and designing. In the same year a new building project was being undertaken at Greenwich: 'At Sir Chr. Wren order to view Spittelfields for Title and to direct Observatory in Greenwich Park for Sir J. More. He promised money.' Later in 1676, he received 3 guineas from Dr Busby for 'the paving Westminster quire', and 5 guineas 'for surveying Law Lodgings'. When Hooke was paid, he could in turn pay his own servants: 'Paid Mary for two quarters 10 sh.', and Harry, his assistant. In spite of irregular pay and Hooke's uncertain temper, Harry seems to have enjoyed his work there. At least he declined an offer of a change when Hooke had the opportunity of exercising a little patronage: 'Sir Chr. Wren's Frenchman dying I preferred his place to Harry—he refused it.'

If Hooke had received what was owed to him, he would have been a wealthy man. He cast his accounts at the end of 1676 and found that the salary that was owed to him for various jobs amounted to £1300. His debts came to 7 pounds 4 shillings, and among the regular items that he had bought, such as coal and wine and shoes, he owed 20 shillings for a quadrant. Anxiety over money never seems to have prevented Hooke from buying instruments and, even more often, tools to make his own instruments. His diaries are a valuable source of information about prices. On 19 March 1794 Lord Brounker signed a bill for expenses incurred by Hooke in his work for the Royal Society.

The items were:

Carriage of quadrant	16 sh
glasses for ,,	20 sh
glasses for telescope	52 sh
frame for quadrant	10 sh

Telescope prices were high but, apparently, there was opportunity for negotiation. On 2 January 1676 Hooke records that Christopher Cock was asking £40 for a 60-foot glass, but was willing to take £15. This may just have been on account, rather than a reduction in price. The usual rate for a telescope of this sort of size would have been in the region of £40, as we can see by the following example of competitive tendering. In 1670 Hevelius wrote to Oldenburg about a telescope of 50 feet that he was ordering from London. He had found Cock's prices lower than Richard Reeve's. Reeve was asking £45 for a 50-foot tube. Hevelius found London prices too high and, in fact, was later forced to decline the offer of a 100-foot tube designed by Cock because he could not afford it.

Smaller telescopes could be found that were a good deal cheaper. Flamsteed mentioned in a letter to Oldenberg in 1673 that he had bought his 7-foot glass for 6 shillings. He was hoping to be able to find one of double length for no more than 12 shillings. At that time, he was completely dependent on his father for money, and was, therefore, most anxious to find the cheapest possible instruments.

Telescopes and quadrants were probably the most expensive items that an astronomer would need to buy, but at least he could hope to find one or two good instruments and use them for some time before the urge for a bigger or better, longer tube or more finely ground lens became too strong to resist. He would need a good pendulum clock with which to count seconds. A clock could cost more than £100, but that expense would not have to be repeated very often. Books, on the other hand, were another matter. Although a great deal of communication was in manuscript form through letters and papers, there was still a great deal that could be learned only from books. This was the case both for the ancient authors and for the more recent works, particularly the catalogues and collections of data such as Hevelius' *Mercurius in Sole Visa* or *Selenographia*. Publication of a new book was an event that was anticipated with eagerness although, unfortunately for the booksellers, the number of actual buyers for scientific books was very small.

In providing books and, to a lesser extent, instruments, the institution of patronage played a major part. In the case of the amateur astronomer, Stephen Gray, who had no surplus from his income as a dyer, contact with someone who could lend him books was of the utmost importance. After he somehow came to know Flamsteed, Gray received from him every year the predictions for the places of Jupiter's satellites. None of Flamsteed's replies to Gray has survived, as far as is known, but he had evidently given Gray reason to hope. Gray's other patron for the loan of books was the Royal Society. From them he borrowed Galileo's and Scheiner's books on sunspots. He returned them some time later with comments on their observations compared with his own. He also pointed out that he could read Latin but had trouble with Italian.

Flamsteed must have been sympathetic. In his early days he, too, had needed to depend on borrowing books. One of the most formative books in his astronomical career had been Sacrobosco's *Spherae* which he had borrowed from a neighbour of his father. Later he borrowed books from his patron Jonas Moore and also from the Royal Society. One letter to

Oldenburg in 1670 shows that Flamsteed did not know about the discovery of Titan because he had not been able to read Huygens' book *Systema Saturnia.* He asked Oldenburg to lend him the Royal Society's copy.

Flamsteed also corresponded with the north country group of astronomers and, particularly, with Richard Towneley, who played in some ways the part of a patron. Here the correspondence involved Flamsteed sending materials like books north to Towneley, and, in return, receiving the status of Towneley's friendship and approval. Moreover, he had also received the micrometer made by Towneley from William Gascoigne's design. In both Flamsteed's main relations with patrons, the tone is more of friendship than humility. Flamsteed was perfectly ready to criticise both Moore and Towneley and tell them where their errors lay.

Stephen Gray, on the other hand, had a relationship with Flamsteed and the Royal Society in which he showed a much greater awareness of his own humble position and dependence. He presents his work to the Royal Society in self-deprecating tones. To Flamsteed he was equally self-effacing. In 1701 he wrote describing a method for finding the latitude and longitude of places on the Moon: 'I know Sir this Method is not accurat enough for you who are furnished with such large and exquisite instruments yet I am apt to believe this may be of good use to seamen . . .' He concluded the letter in a tone that suggests a friendly relationship: 'Sir I am your most Affectionate Humble Servant, Stephen Gray.'

As the relationship progressed, Gray brought himself to ask for Flamsteed's opinions as well as his calculated places. In 1706 Gray had observed the solar eclipses. He sent his data to Flamsteed, together with a description of his method of drawing the meridian line. In return, 'I would be glad to receive a line or two from you with your thought on what I have written and if it be not to great a Trouble some of the Principle Phases of the solar eclipse observed by you will be very wellcome to me.' His desire to possess Flamsteed's *Works* is expressed in Gray's most enthusiastic strain. He must have been fairly pleased with the elegance of the sentence that he turned: 'Sir I hear by a Gentleman that came lately to see me that your works are printing he haveing seen some sheets of them printed witch I am very Glad to hear of haveing a longeing expectation to see them though I doubt they will be to Great a Treasure for me to think of procureing except the care you were once pleased to say you would take that such as I should come by them easy enough be greater then I deserve who am the meanest of those being Sir, Your most Humble Servant.'

Flamsteed, of course, found the correspondence with Gray useful. Not only did he receive data for his work on Jupiter's satellites, but also eclipse data that he considered good enough to send to the Royal Society for publication in the *Philosophical Transactions* along with his own. In the publication process, another function for the patron emerges. Not only did he bring his protégé to the notice of London scientists but, also, he literally 'protected' him in a dispute. In 1715 Gray travelled to the house of another of his patrons, Mr John Godfrey, near Faversham in Kent. There he was to assist Dr John Harris. Harris was a notorious pluralist with a record-breaking number of parishes in Kent and Sussex that financed his dabbling in science. He grossly mismanaged his parochial business and his finances to the point where he died as a pauper at Godfrey's house.

In 1715, however, Harris had decided to observe the eclipse, and he asked Gray to assist him. He then sent the results to the Royal Society, claiming all to be his work except for an error of a whole minute in the times which, when it was pointed out to him, he attributed to Gray. Flamsteed came to Gray's rescue, asserting that 'This I am apt to believe Mr. Gray will not owne to be his fault, he was to cautious to commit such an one but the Dr who owns himself not so expert must bear the blame for this not being all to rights.'

Patronage was a system taken very much for granted in the seventeenth century. Anyone was likely to accept favours or grant them to others when position and state of mind warranted it. Thus, Jonas Moore provided Flamsteed with lodgings, books, instruments, clocks and a servant, when his position allowed him to do so. He also introduced Flamsteed to the court and won him favourable opinions from the King. In his own early days, Moore himself had been made tutor to the Duke of York by a favour from Charles I. In the Moore–Flamsteed patronage system, Flamsteed's ability to provide Moore with good brown ale from his father's brewery was a factor in Flamsteed's favour. The system was so well established that often it worked without causing rancour or ill feeling. John Pell, who had not been entirely happy with the position given him by his patron the Archbishop, nevertheless wrote a passage in his *Idea of Mathematics* asking for more patronage for science and mathematics.

Flamsteed was inclined to have mixed feelings about Moore as a patron. Although it is true that Moore's efforts to supply a quadrant came to grief with Hooke's design (see chapters 4 and 7), nevertheless he had, before he died in 1679, provided Flamsteed with various lifelines. One was the rather desultory interest of the King. Thus, after Moore's death, Flamsteed was able to present himself at Court with a newly calculated tide table as proof of his practical usefulness, to ensure that the Observatory would be allowed to continue. According to a letter that he wrote to Seth Ward, he pointed out to the King the importance of the Observatory to the country's prestige abroad, as well as to the practical results that might be expected if an adequate salary were to be paid.

Moore had also introduced Flamsteed to Lord North, and this proved a most valuable gift, as it was North who, in 1684, presented Flamsteed with the living of Burstow and so contributed to keeping the Observatory solvent.

The favour of the King was not easy to win in any reign. Flamsteed, as the King's Observator, had inevitably to try to keep the royal goodwill, although while Moore was alive, he had left promotion in his patron's hands. Others were more eager to try to obtain some kind of favour for themselves. Thomas Street made a name for himself as astronomer and astrologer by publishing his *Astronomia Carolina* in 1661. He 'presented it well bound to King Charles II and also presented it well bound to Prince Rupert and the Duke of Monmouth but never had a farthing of any of them', according to Aubrey. His lack of success as a courtier is understandable in the light of Aubrey's anecdote: 'He was of a rough and cholerique humour. Discoursing with Prince Rupert his highness affirmed something that was not according to Art: says Mr. Street, "Whoever affirms that is no mathematician". So they would point at him afterwards at Court and say "there's the man that huff'd Prince Rupert".' Offerings to royalty were not usually lucrative. Halley had little reward for his letter summarising the *Principia* that he presented to King James. In fact, Halley spent a great deal of money on the *Principia*, as he bore all the printing costs himself.

Looking at astronomers as a group shows that they were not generally prosperous men. A few suffered from serious poverty. A similar small number enjoyed the luxury of a high standard of living from which they could easily afford the expenses of astronomy. In between, the great majority earned a living which just barely allowed a surplus for astronomy. The Church and the universities were the largest providers of funds. An interesting side issue here is that astronomy itself provided very little money. A few astronomers sold almanacs but no-one made much profit from books. Teaching, then as now, was one of the few ways of making money from an academic subject. Calculators and labourers could make a little money from assisting others. Their wages depended on their employers' sources of income. After all, the Astronomer Royal's £100 per year (£90 after tax) was the only professional salary earned purely from the practice of astronomy.

6

Communication

When a group of scientists shares a common pool of assumptions about their subject, its truths and its problems, they can be said to form a thought collective (Fleck 1979). Members of such a group will tend to be working on similar problems because they have similar views of what constitutes knowledge and what is an important but soluble problem on its fringes, but belonging to a thought collective certainly does not imply agreeing with one another. The scientists working in astronomy in Newton's time present an intricate series of interconnecting and overlapping communication systems. A diagram of these systems would show many stops and dead ends where individuals refused communication or collaboration. Yet they shared a common view of which problems were interesting and soluble and also a common view of what tests might be applied to a potential solution.

Communication was not always a blessing and then, as now, there were times when men were not anxious to publish, or even willing to write letters. Newton went through several periods when he was reluctant to write at all. Flamsteed, on the other hand, had extremely bad relationships with many of his contemporaries and yet kept records of a large volume of correspondence that continued throughout his working life.

One way of looking at the function of communication is to assume that it goes beyond the giving and receiving of information. It also preserves each individual in his place in the community, allowing him to share in the prevailing assumptions and concerns of that particular group. The amount and effectiveness of communication that appears can take different aspects depending on the axis examined. The means used to communicate will also vary: speech, letters, papers, journals, even diaries will be relevant. Individuals can be assessed for the extent to which they wished to communicate at all. Some might have wished to keep their own ideas secret but learn as much as possible from others.

The amount of private and personal information content can be examined especially in letters although, unfortunately, many letters that survived by being published have had their personal content removed by ruthless editors. The extent to which astronomy interacted with other sciences can also be inferred from surviving records of communication. One of the most interesting conclusions reached from searching the known correspondence is that, in communication by letters in particular, certain men were key figures who facilitated interaction among others. Although Flamsteed is the most important of these for astronomy, there were others who were also helpful. Probably the most surprising of these is John Collins, a tireless correspondent whose own contributions to science were connected with fisheries and the keeping of accounts; but he was a man who was able to facilitate the work of others on all subjects.

Flamsteed, in spite of his reputation for solitariness, kept correspondence on astronomical subjects going at one time or other with about 75 of his contemporaries (see table 6.1). A large proportion of the 76 volumes of his papers preserved at the Royal Greenwich Observatory consists of letters. This volume seems all the greater when Flamsteed is compared with Newton, who, in some years, had little continuing correspondence, and in many cases only a single letter was involved (see table 6.2). Some of the subjects show the peculiarity of Newton's state of mind at the time.

Secretaries of the Royal Society were, by their position, required to make some effort at corresponding with the philosophically inclined, both in Britain and abroad. Henry Oldenburg, Hans Sloane and Edmond Halley were the most conscientious in carrying out this task. Henry Oldenburg not only replied to Flamsteed's early letters from Derbyshire, but actually invited him to London and introduced him to Jonas Moore, among others. Halley, when he came to office, wrote to William Molyneux and George Ashe in Dublin, saying that the Royal Society was eager to establish a correspondence with the Dublin Society. For his part, he promised that replies would be prompt, a condition that had apparently not operated in the past. The position of Secretary did not bring high status with it. Halley was ordered to sit at the lower end of the table at meetings and he was not to wear a wig.

Although letters form by far the largest portion of the known communication, some astronomers left other messages for posterity. Flamsteed, unlike Newton, was sufficiently concerned with his own image to write several instalments of an autobiography. He seems to have intended that we should read it, and was happy to include such personal details as

Table 6.1 Flamsteed's correspondents on astronomical and related subjects.

George Ashe	Isaac Hawkins	Henry Sherwin
William Bosseley	Thomas Heathcott	Thomas Smith
Thomas Brattle	Thomas Hill	Henry Stanyan
Richard Burnet	Henry Jacob	Henry Thomas
Mr Burtsal	Luke Leigh	Francis Thompson
John Caswell	J Lowthorp	Stephen Thornton
John Collins	J Machin	Richard Towneley
William Derham	Samuel Molyneux	Richard Waller
William Ella	William Molyneux	John Wallis
Nicholas Fatio	Jonas Moore	R Walsh
Mr Glen	Isaac Newton	Seth Ward
Stephen Gray	John Pell	William Whiston
Henry Greatorex	James Pound	John Witty
David Gregory	Matthew Randall	Sir John Worden
James Gregory	Joseph Raphson	Christopher Wren
Edmond Halley	Abraham Sharp	Matthew Wright
John Harris	Edward Sherbourne	Richard Wroe
Imanuel Halton	Thomas Sherlock	George Young

Correspondents abroad

Christoph Arnold	Jean Dominique Cassini	Gottfried Leibnitz
Fr Balsamus	Chardellou	Olaus Roemer
Francisco Blanchinus	John Baptista Ciampini	J Spangenburg
Charles Boucher	Johannes Hevelius	Weihelius
I Boulliau	Gottfried Kirk	J J Zimmermann

Table 6.2 Newton's correspondents 1676–94.

1676–87			
Aston	1	Flamsteed	17
Aubrey	1	Halley	7
Boyle	1	Hooke	5
Briggs	1	Lucas	1
Burnet	1	Maddock	1
Collins	2	North	1
Crompton	1	Oldenburg	13
1688–94			
Barton	1	Harris	1
Bentley	3	Huygens	2
Charlett	1	Menicke	1
Corke	13	Pepys	4
Covel	13	Wallis	2
Gregory	2		

descriptions of his childhood illness. Probably he saw its main purpose as setting straight the record of his dispute with Newton. The rest of his life was peripheral to this great concern. His marriage, for instance, receives scarcely a mention.

Newton, whose position in the quarrel was better known to most contemporaries and has received more attention ever since than Flamsteed's, was a man who never apparently enjoyed contact or communication of any sort. He was regarded with awe by those of his contemporaries who knew anything at all about his work. His peculiarities were accepted and tolerated, even expected after the publication of the *Principia.* Those who understood it assured the world that it was the most important scientific book that any of them was likely to live to see. With this reputation, Newton spent the 1690s going through his most disturbed period, refusing to communicate with anyone except on the occasions when he sent letters so peculiar that the recipients must have found replying difficult. Accusing Locke of trying to embroil him with women brought only the gentle reproach of one who was 'intirely and sincerely your friend'. Newton was sometimes aware that his behaviour was exaggerated. In the case of Locke he wrote and explained that he had been feeling ill from sleeping 'too often by the fire' (Westfall 1981).

When Flamsteed and Newton tried to work together, the resulting clashes resounded for many years and were not easily to be softened. Yet even these two were not always at loggerheads. In the 1690s they were writing to each other in terms of friendship. There is even an expectation of fellow feeling in Newton's comment to Flamsteed in a letter of 1691: they both share the problems of working in a science with the long baseline of astronomy: 'I would willingly have your observations of Jupiter and Saturn for the next 4 or 5 years at least before I think further of their theory but I had rather have them for the next 12 or 15 years. If you and I live not long enough Mr Gregory and Mr Halley are young men' (10 August 1691: Turnbull 1961).

Flamsteed, at this time, had not yet quarrelled with Gregory. In fact he had not yet met him in person, but this reference to Halley shows either that Newton was not bothering to be tactful or that he did not know that Flamsteed had already found much fault with that young man. Newton himself liked Halley, whose attitude to him was always respectful. In spite of this false step Flamsteed still regarded Newton as a friend in those days. Halley was the enemy whom Flamsteed would not even mention by name. In a letter of February 1692, he wrote to Newton answering what he took to be Halley's suggestion that it was time he published something. He was not including Newton in his anger against Halley: 'I take your advice very kindly because I know you are sincere in it and wish me all the success I can desire in my labours and I shall give you very substantial reasons why I cannot do this at present' (24 February 1692).

The letter is long and gives Newton a detailed, but still friendly, explanation of Flamsteed's determination to publish a complete catalogue. The condition is that it shall be when Flamsteed is satisfied that the data are ready to be made public.

Flamsteed obviously felt pressurised by the scientific community as a whole, but was determined that no-one should see his observations before he was ready. Newton's letters to Flamsteed in the 1690s constantly reassure Flamsteed that he had not made and would not make public the observations that were being sent. He promised that he would not even let Halley see them although, as he pointed out several times, he was constantly hoping that the rift between Halley and Flamsteed would be healed.

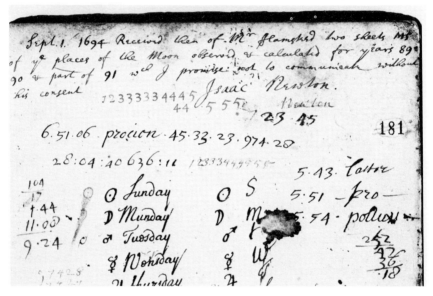

Figure 6.1 Isaac Newton's signed receipt for the places of the Moon given to him by Flamsteed. Here he gives his promise not to communicate them to anyone without Flamsteed's permission. Below, a pupil practises astronomical shorthand. Royal Greenwich Observatory.

Soon, of course, a rift began to develop between Newton himself and the undoubtedly difficult Flamsteed. In January 1699, Flamsteed annoyed Newton by mentioning in a letter to Wallis intended for publication that Newton was about to solve the lunar problem, a result that would be pleasing to Flamsteed, whose data were being used. Newton himself knew that such a solution would take time and was certainly not imminent. He pounced angrily on Flamsteed for what seemed a very premature comment. '. . . there may be cases where your friends should not be published without their leave and therefore I hope that you will so order the matter that I may not on this occasion be brought upon the stage' (6 January 1689). This was the man who, four years later, dragged Flamsteed upon the stage without hesitation. Since it did not produce sympathy, the incident may have provoked a desire for revenge. A little desire for revenge would certainly explain Newton's later behaviour. When Newton had manoeuvred Flamsteed into agreeing to have his catalogues published as the *Historia Coelestis* he interfered in what seemed to Flamsteed a very arbitrary way. Without explanation, Newton on occasions commanded the printer to stop the press and reduced very slow progress to complete standstill. The whole affair is evidence of a breakdown of communications on a large scale.

In some ways Flamsteed has the characteristics of the tragic hero. Certainly he brought disasters on himself by his pride and inability to predict or judge the effects of his words and actions on others. The moment when he lost control of the publication of his work can be clearly seen. Stirred to action by accusations from Halley and Gregory that he had produced very little in his time at Greenwich, Flamsteed wrote a plan for his *Historia Coelestis* in which, by estimating the number of pages that he would fill, he showed how large a volume of work he had already done. In his plan he mapped out the contents of the various books, including his catalogue of stars which he had not, in fact, finished.

With this plan Flamsteed sent his nephew by marriage, James Hodgson, to London, instructing him to show it to a friend who would, he knew, do him justice at the Royal Society. Unfortunately, Hodgson was talking at one end of a busy crowded room when he saw the person for whom the draft was intended. He handed it to someone standing in between with a request to pass it on. In the noise of the room, the request was misheard, and the document was given to Sloane, the Secretary. Sloane found it so interesting that he read it aloud to the assembled members. The Society as a whole was enthusiastic for publication. In this way Newton came upon exactly what he wanted: the possibility of having all Flamsteed's observations available to him whenever he wanted them. From then on, Flamsteed was never leading from in front, but always being pushed from behind.

At about this time, Newton was considering a new edition of his *Principia*. If he could solve the lunar problem, how satisfying to be able to include it in the new edition: undoubtedly Flamsteed would have more observations of the Moon than Newton had yet seen. With those in his hands he might yet make a coherent theory of the Moon. He went down to Greenwich in good spirits, ready to be gracious with his good news. His Highness Prince George, husband of Queen Anne, had agreed to finance the publication under the auspices of the Royal Society.

Thus, from Newton's point of view, the dispute with Flamsteed began only because Flamsteed was slow to appreciate his chance to make public the results that he had obtained in his thirty years as Astronomer Royal. Through the long years of acrimonious argument and sniping that went on between Newton and Flamsteed, Newton, at least, was clear about his objectives. His ill treatment of Flamsteed must have seemed to him like just punishment for Flamsteed's stubbornly uncooperative behaviour.

Seen from the Observatory at Greenwich, the quarrel must have looked very different. For one thing, the advantages all seemed to be on the other side. Newton was already Master of the Mint when he was elected President of the Royal Society in 1703.

Flamsteed's attitude to the plans for publication was at first accepting, even cautiously enthusiastic. Letters to his own friends and admirers show that he was willing at first to cooperate, although, of course, he expected authorial rights to proofread and make corrections. He also expected to be paid something out of Prince George's fairly generous allocation for expenses. Newton's behaviour, on the other hand, was to Flamsteed quite inexplicable. He wrote to Sharp: 'Sir I Newton hindered its progress by continual shuffles and tricks' (18 June 1719). Whatever the reasons on any given occasion, the printing was certainly very slow. Only 98 sheets had been printed after four years.

Newton was even more difficult about the payment of Flamsteed's expenses, although the committee appointed to administer the Prince's money had agreed to pay the cost of hiring calculators. For his part, Flamsteed kept a scrupulous record of his expenses. For example: 'Went to London. Expenses 3/6'. A carefully made up account of major expenses such as salaries but excluding the small travelling expenses came to £173. This was presented to Newton in 1705. After considerable delay he paid Flamsteed £125 and returned £25 to the Prince's administrators. No explanation was given for denying the money to Flamsteed, whose natural reaction was to think that not only was he very far from being rewarded for all his work, he was not even being justly treated according to the terms of the Prince's offer. Clearly, any reward would have to come from the satisfaction of knowing that future generations would see and use his work. On this Flamsteed concentrated all his attention.

Although there is no need for posterity to share out blame, several interesting comments can be made on the dispute. First, Newton and Flamsteed both wasted opportunities. At their most acrimonious, not only did they fail to collaborate, but they actually tried to hinder each other's achievements. Newton certainly tried to limit Flamsteed's activities to the areas that would provide useful data. In 1704 Flamsteed stated to the committee on publication that he needed money so that he could hire calculators to help him compute the places of the remaining stars and to make corrections etc in other data. The money that was allocated to him was specifically assigned to the calculation of places in the solar system,

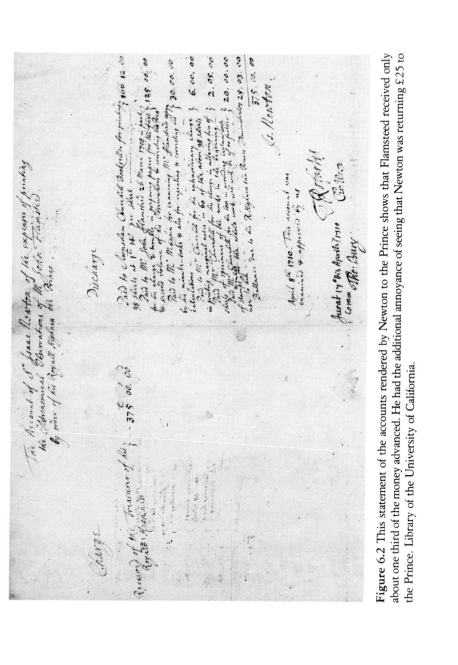

Figure 6.2 This statement of the accounts rendered by Newton to the Prince shows that Flamsteed received only about one third of the money advanced. He had the additional annoyance of seeing that Newton was returning £25 to the Prince. Library of the University of California.

ignoring altogether the need to work on finishing the star catalogue. Newton's interest at the time lay exclusively in the solar system. Flamsteed was not a man to be deflected from his course and he finished his star catalogue anyway, with his own money.

Second, the dispute does not show that Flamsteed was reluctant to communicate. His reluctance was over communicating material before he was ready and before it was in a form that satisfied him. Whether Newton, on the other hand, might have done more if the raw data had been given to him in a steady stream straight from the Observatory is an interesting, but not very useful, speculation. He never achieved a satisfactory lunar theory, and became more and more disenchanted with the whole subject, realising the immense difficulty of solving the problem with available data. Flamsteed's slowness in divulging data was probably not significant.

The story of the publication of the *Principia* is an example of a very different process from the *Historia Coelestis.* The book had its origin in two personal visits paid to Newton in Cambridge by Halley. On the first of these, Newton told Halley that he could demonstrate a mathematical proof that the law invented by Kepler for the speed of motion of the planets was correct. He couldn't at the moment find the proof, but promised to send it to Halley in London when he unearthed it from his papers. He did find it and duly sent it to Halley, who went straight back to Cambridge to see what Newton was on to. At this visit, Halley began to see the scope of Newton's originality and the value of his synthesis. He suggested that Newton should write a book before he lost the papers again.

In this way a collaboration of immense value to science and to the whole climate of thought was begun. Halley took on all the expense and all the worries of publication. Newton was troubled by nothing except the writing and the dispute raised with Hooke, who claimed to have proved the inverse square law before, but couldn't find anyone to corroborate his claim. This dispute caused very poor relations between Newton and Hooke, but, as they had never attempted to cooperate, and as Hooke died in 1703, the effects were less visible than the protracted animosity between Newton and Flamsteed. Hooke's main achievements were, in any case, in fields other than astronomy.

Charles II had agreed to the appointment of a royal astronomer on the understanding that he was to improve navigation. By the end of the century, no major advance had been made in finding longitude at sea. Flamsteed could have pointed to his steadily increasing store of data as the best chance there was. He worked particularly hard on collecting observations of the moons of Jupiter in the hope that accurate times for their appearance and disappearance would enable mariners to fix their time, as no-one had yet invented a reliable marine chronometer.

In his capacity as Astronomer Royal, with this aim at least partly in view, Flamsteed was very free in giving as well as receiving information, particularly on Jupiter's satellites. He supplied his tables for the predicted places of the satellites each year to many amateurs around the country, and he also gave advice and comment on the issues of the time in astronomy. Even so, this openness had its limits. Flamsteed clearly instructed his correspondents that they were not to pass on any information from him to any third person without his approval. Frequently, assurances in letters to him indicate how strongly he impressed this on his correspondents.

The tables of motions of Jupiter's satellites led Flamsteed into many correspondences as word spread that they were a main focus of research for navigation. Seamen like Henry Thomas and Henry Stanyan were obviously very interested in the practical improvement of navigation.

Others, like Thomas Brattle of Boston, New England, and George Young of Bere Regis, were anxious to embark on a correspondence for their own intellectual satisfaction. Both of these expressed their sense of isolation and lack of congenial companions with whom to talk astronomy. One factor common to correspondents is that they found Flamsteed open and helpful. He responded generously to requests for advice and information, even when not clearly related to his job.

George Young opened a correspondence by asking for help with choosing and buying books. In the small town of Bere Regis he found very few books available to him 'for astronomy the delight of Princes is here the contempt of Rusticks' (4 October 1708). He feared that London booksellers would cheat him when they found him to be a naïve country bumpkin. He therefore asked Flamsteed's opinion of a fair price for useful astronomy books.

Flamsteed's response can be inferred from the humble thanks 'for your offer and advice' in Young's next letter. Moreover, Young escalates as the years go by through gratitude and thanks to genuine-sounding concern for Flamsteed's welfare, particularly his health. Rather tactlessly, he adds that however that may be and whether he gets better or not, Flamsteed's memory will always be held dear. In his last known letter, written in 1715, just four years before Flamsteed's death, Young sounds genuinely concerned about the demands he has made on Flamsteed's time: 'I can't express my gratitude and sense of ye obligacion but am really concerned that you should take so much pain and trouble for me.'

In return for advice and gifts such as the tables sent, Flamsteed received some eclipse data and much discussion of the problem of the Moon's orbit. In addition, and probably of great importance to Flamsteed, he had a wholly devoted partisan on the great matter of the publication. Letters from men like Young who were uncontaminated by close contact with Halley or Newton must have been a great consolation to Flamsteed, and they alter still further the picture of a lonely, isolated misanthrope.

Thomas Brattle in New England also found Flamsteed helpful. Not surprisingly, one of Brattle's main difficulties was with the practical problems of communication. Letters just didn't flow smoothly from Greenwich to New England. More surprisingly, perhaps, enough letters did reach their destination for a correspondence to subsist. The dates of the letters mark the slow progress of the relationship.

Brattle first wrote to Flamsteed in 1692. He sent two letters that year and then waited eleven years for a reply. In 1703 he was informed by a friend from England that Flamsteed had received only one of the letters, but had actually written and sent off a reply to that one. The reply had never reached Brattle. In spite of this rather patchy beginning Brattle still felt enthusiastic enough to write to Flamsteed again at once. In 1703 the correspondence at last got off the ground.

Letters seem to have taken 5 months, on average, to pass from hand to hand. One, in particular, is damaged by salt water, but all are still legible. Brattle seems to have expected that letters would arrive safely and when, in 1707, he had not heard from Flamsteed since 1704, he assumed that Flamsteed had not written, rather than that a letter had gone astray.

None of these correspondents was likely to write very personally. Self disclosure must not have seemed appropriate, since the purpose of the letters in all cases was to give and receive academic or practical information. Brattle, living in the New World, makes no assumptions that Flamsteed would be interested in his experiences. Instead, he keeps to the matter in hand, with the merest incidental references to the world around him.

Henry Stanyan comes as close to a personal subject matter in his correspondence with Flamsteed as anyone. He was a former pupil, but had an irrepressibly unacademic tone. Stanyan went off to be tutor to a family in Berne in Switzerland. He wrote cheerful, chatty letters, full of descriptions of the Alps, with just the occasional nugget of eclipse data to offer to his old master. He was obviously on the best of terms with Margaret Flamsteed. Presumably she had not resented his apparent familiarity.

The correspondents considered so far wrote to Flamsteed to give information as well as to receive it, although the value to Flamsteed naturally varied greatly. As an author and practising scientist he also had correspondents who wrote to question or criticise his work. What we know of Flamsteed might lead to the suspicion that he would not receive such letters graciously. In fact, there is at least one case in which Flamsteed did accept criticism and in doing so won another admiring partisan. The criticism was, however, expressed in the form of questions and was not aggressive. Second, the author was not well known in scientific circles. No comment that he could make could have stung like the words of the up and coming young Halley who was accepted and approved even by Newton.

The man who entered into this relationship with Flamsteed was a sea captain named Henry Thomas. When he first wrote to Flamsteed he was lieutenant of the *Humber.* Thomas asked intelligent and probing questions, but clothed them in a little praise and considerable respect. He was not an effusive writer. The first letter mentions 'that accurate part of your labour', the *Doctrine of the Sphere*, but asks for more detail about Flamsteed's description of the inclination of the Moon's orbit to the ecliptic. Flamsteed was ready with an explanation. Thomas is quite right in his remarks. The text should have read 'left' instead of 'right'. Thomas' reply, trying hard not to be offensive, succeeds in being confusing. He expresses his thanks 'as I ought' for Flamsteed's satisfactory explanation and proceeds to tie the reader up in knots about how an error could have crept into the work of one so careful in proofreading as Flamsteed must undoubtedly be. He has another awkward question: how can it be that the tables published in Whiston's book of 1707 and called 'Flamstedianae Correctae' are so different from Flamsteed's previous figures? Making the point that Flamsteed might have been wrong the first time, when he published the tables, seems over-bold, and Thomas feels that it calls for some sort of apology. He stumbles through the rest of the letter in would-be elegant compliments.

The last letter from Thomas, written on board his ship at Spithead in September 1711, is more in keeping with his character. In it he gives a straightforward factual account of the secret expedition of the fleet to Canada, where an attempt was to be made to capture Quebec and Newfoundland from the French. He points out drily that although Flamsteed's last letter full of valuable information did not arrive in time for the voyage and cannot, therefore, be said to have been of any help, yet he feels that he ought to express his thanks anyway. The rest of the letter passes off without compliment until the end, where he declines to keep Flamsteed away from the public service any longer, and suggests that he may do himself the honour of visiting Greenwich some day. There is no suggestion anywhere of Flamsteed having taken umbrage.

The data that Flamsteed received from correspondents were usually patchy and of variable accuracy. From Stephen Gray of Canterbury came data which were likely to be accurate, although bad weather and personal problems interfered with the consistency of the observations. Gray was one of the most painstaking of the observers. Right from the beginning of his known correspondence there is evidence of his careful, methodical approach. Gray was certainly a scientist in his own right, and his correspondence is valuable beyond the relationship with Flamsteed. In fact, not all of it was addressed to Flamsteed. Much was addressed to the Royal Society through the current Secretary. Gray's first steps towards scientific recognition began with a letter to the Royal Society written in 1696, when he was thirty. Since he already knew Henry Hunt, he addressed his earliest letters to him. The first letters were on a variety of subjects, including chemistry and biology. Gray showed from the beginning an interest in technical problems and an ability to solve them. He was able to apply his interest to grinding lenses and making telescopes. For the moment, his letters on water microscopes and, later, on grinding specula of a nearly parabolic shape must have appealed to the medically-minded Sloane who was then Secretary and Editor of the *Philosophical Transactions.* Gray's success seemed likely from the very beginning when his first known letter to Henry Hunt was published in the *Philosophical Transactions* of June to August 1696.

Between 1696 and 1703 Gray wrote more than 20 letters to the Royal Society, addressing later ones to Sloane directly. Some of these letters were published and some were read to the Society. One that had strong contemporary interest was an experiment to discover which of variously shaped and sized hour glasses filled with water and sand would provide the best chance of an accurate time keeper for ships at sea. This letter aroused so much interest that Gray was asked to provide more details for the Society. The interest was mainly academic and the idea was not used to revolutionise navigation.

Nevertheless, Gray was well on the road to recognition. He was apparently accepted as scientific correspondent in Canterbury. When an interesting problem needed further investigation, Gray's name immediately sprang to mind. In 1700 the Journal Book presents an odd discussion going on at the regular meeting: 'There was presented from Mr Hunt of Canterbury a piece of pork taken from a well in chalk near Maidstone where it had layn many years. He was thanked for it and Mr Hunt was ordered to get a candle against meeting to try if it would burn or not' (13 September 1700). Apparently the pork was sent by someone called Gray. Sloane wrote and asked for more details, eliciting a somewhat surprised reply: 'I have received your kind letter wherein you desire a more particular account of the Pork I sent the Royal Society. This is therefore to certify you that I know nothing of the matter having never till now heard that any such thing was done. The person that sent it is either of my name or one that for I know not what reason personates me.' The frosty indignation of this letter indicated that Gray certainly didn't see the funny side of someone sending a piece of pork to the Royal Society in his name. He was similarly unenthusiastic when designated to look into the Veal–Bargrave affair (see chapter 7). He did what he could because, that time, it was Flamsteed who asked him, but being in the news was not really to his taste.

In spite of these unwelcome diversions, Gray began to feel more settled in the scientific community. He began sending astronomical work to Flamsteed. In 1699 he sent a table of eclipse data (as far as we know this was the first). In 1703 he sent at least two letters, now lost, to the Royal Society. We know that he sent them because there were references to them in the *Philosophical Transactions.* These letters were both on the subject of sunspots. Apparently Sir John Hoskins, who was then President of the Royal Society, took an interest in Gray's work in this field. He asked Gray to make observations whereby the rate of the Sun's revolution on its axis might be determined. Gray responded in two long letters on sunspots with some careful observation and impressively relevant speculation on the

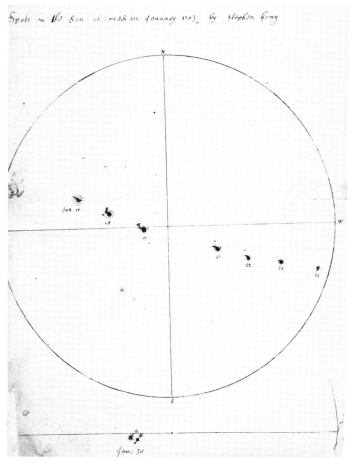

Figure 6.3 One of Gray's careful drawings of the progress of a spot in the Sun in 1703. These drawings are still of interest to astronomers as they show the activity of the Sun at the time when it was just emerging from the period of low activity known as the 'Maunder Minimum' (Clark and Murdin 1979). Royal Greenwich Observatory.

nature and movement of the spots, and on the rotation of the Sun. However, neither of these letters was published.

In the year 1703, when Flamsteed began his major dispute with Newton and Newton himself became President of the Royal Society, Gray's star went into eclipse. No more letters of his were published while Newton was President, with the exception of one in 1720 on electricity. Yet Gray's scientific interests continued to develop.

The sudden end to publication from Gray has led to the question: did Newton deliberately suppress Gray's work and, if so, was he seeing Gray as some kind of protégé of Flamsteed? After 1703, Gray seems to have turned his attention steadily to astronomy, which led him to write at least twenty letters to Flamsteed, mainly on astronomical subjects. In these letters he showed strong loyalty, both to the validity of Flamsteed's opinions and to Flamsteed personally in the publication dispute. In the first of Gray's unpublished sunspot letters he referred to Flamsteed as: 'that most accurate Astronomer my Honoured Friend Mr John Flamsteed in a letter which I had the honour to receive from him . . .' (3 April 1704). Such phrases might have been remembered with rancour by Newton in the days when his anger against Flamsteed was high.

Gray was never in any doubt about where his loyalty lay. In 1707 he wrote mentioning the *Historia*, hoping that Flamsteed would 'Reap some Benefit of your Indefatigable Labours besides that Eternal Commendation that will be allwais due to your Memory for notwithstanding the opposition and aspersions you have met with from Ignorant Malitious or unjust detractors whoe may have Indeavoured to Depretiate your performance, Posterity will know how to value them' (26 January 1707).

Gray's dismal publication record in the early 1700s may have depressed him at the time, but it certainly did not finish him off. He entered a new period of success and recognition in the 1730s when Newton was dead and Sloane had become President. Gray himself was sixty but he was at last free to consolidate and add to his earlier work on electricity. By this time, electricity had become so fashionable that ladies and gentlemen unconnected with the Royal Society came to watch Gray produce sparks and convulsions with animals and sometimes people. In his enjoyment of the new-found acclaim, Gray seems to have abandoned astronomy.

Letters were certainly an important medium of communication. Meetings both public and private, though more difficult for us to assess, played an even bigger part. Gray had worked in astronomy in the early part of the century, far longer than the letters to Flamsteed indicate. In 1712 he left Canterbury and went to live in London. From then on the Journal Books of the Royal Society indicate that he was frequently present at meetings: 'March 1st 1714/15 Mr Stephen Gray had leave to be present at the meeting.' He was present many times. He became bolder and on 1 November 1716 was found showing 'a Quadrant with a new Projection useful for taking the Distance of the Sun's Spots from the Eastern or Northern limb of the Sun'. He began to present papers at meetings, at first always in conjunction with someone else. On 6 December 1716 'Mr Gray's account was read of the same eclipse agreeing with that of Dr Foulkes.' On 7 February 1717, Mr Desaguliers read a paper of observations by Mr Stephen Gray and himself. At last, though, on 28 March 1717: 'Mr Stephen Gray delivered a paper of Observations of the Hugenian Satellite of Saturn . . .'

In Desaguliers Gray had found a temporary collaborator. Together they made observations at Desaguliers' house on Westminster Bridge. When Flamsteed died in 1719 Gray was not left alone but was well established in London and his results were published, sometimes with those of Desaguliers, sometimes independently. Although they form one of the few examples of long-term collaboration between astronomers in this period, Gray and Desaguliers seem not to have become friends, as it was Desaguliers who later made the remark about Gray's temper not allowing anyone else to work on electricity. This is probably not surprising as we have already mentioned Gray's difficulties in personal relations. His collaboration with Dr Harris in observing the total eclipse of the Sun in 1715 is further evidence that trouble accompanied Gray like Flamsteed and Newton themselves. He would have been a difficult man with whom to work.

Is there, then, enough evidence to suggest that Newton might have been actively hostile to Gray? Newton, as we know, was the man most likely to have done all in his power to encourage the publication of astronomical observations, particularly of the phenomena in which he was most interested: the Moon and the solar system. His quarrel with Flamsteed was over what seemed to Newton to be Flamsteed's suppression of data. Not surprisingly, Gray was not discouraged from producing data, even at meetings where Newton himself was present. The rejected sunspot letters of 1703–4 are much more than a straightforward reporting of data and they were possibly seen in a different light: an unknown from Canterbury putting forward theories of solar rotation could well be suspected of incompetence.

Later, the 1708 letter on electricity showed, from Newton's point of view, that Gray had diverged from astronomy into an area which he felt was a fringe investigation hardly worthy of serious scientific thought. In this capacity, Gray, as a friend of Flamsteed, was certainly of little interest or use to Newton, and his work could well be ignored.

Gray's greatest success came from work done on his own or with an aristocratic assistant like Granville Wheler or John Godfrey, neither of whom seems to have contributed much to the thought of the experiments but everything to the space and equipment. In 1732, when his work on electricity was well known to contemporaries, Gray was awarded the

Copley Medal of the Royal Society and was made a Fellow. In the long run, his correspondence with Flamsteed can be said to have helped him to clarify his thoughts and to keep him in the main stream of one tributary of British astronomy. His collaborations and his success after Flamsteed's death show that he was not dependent on Flamsteed.

One correspondent who is known best through his relationship with Flamsteed is Abraham Sharp. Sharp played an important part in constructing the mural arc for the Royal Observatory. He was a skilled instrument maker by the time he worked for Flamsteed and he was mainly responsible for the extremely painstaking task of engraving the measurements on the limb of the arc, a job which took him fourteen months. Sharp has left a distinct image of his personality in his exquisitely written volume of letters to Flamsteed after returning to Yorkshire.

Life there seemed to Sharp to be a form of exile, and much of his mental life went into his letters. The letters show a two-way flow of information and assistance from 1702 until Flamsteed's death. When Sharp initiated the correspondence by a letter written to Flamsteed on 2 February 1702 he was anticipating a fairly high level of exchange of ideas and objects. He had already made himself a 6-foot telescope, although he complained of the lack of facilities in his remote home. He offered to assist Flamsteed in any way possible. Sharp would be glad 'to be serviceable', and behind the offer is the appeal that by maintaining a working relationship with him Flamsteed will preserve him as a member of the scientific community.

Even in Yorkshire Sharp heard that Flamsteed's catalogue was to be published soon. A 'friend of mine lately in London' heard that the work was in hand. Sharp, like Stephen Gray, was hoping that he might be given a copy. Whether or not this sort of reminder pleased Flamsteed it obviously proceeded from a wholly genuine desire to possess the information, as well as to please Flamsteed by the flattering eagerness of his friends and assistants. Sharp and Gray mark the geographical extent of Flamsteed's operations in England and they were both kept well informed by letters and by keeping careful track of travellers. Both travelled themselves, although Sharp did so less and less as he grew older.

Travelling and visiting each other was one of the possible methods of communication least used by the astronomers of the time. Flamsteed avoided leaving Greenwich if he possibly could. Newton stayed firmly at Cambridge, but Halley, who was always much more mobile, came to see him with the questions that set in motion the writing of the *Principia*. Later, when Newton moved to London, he occasionally visited Greenwich, to the annoyance of Flamsteed, but clearly preferred to stay at home. He made no effort to return to Cambridge.

London was felt to be the centre for astronomy and for instrument making much more than Oxford or Cambridge. There is little doubt that the Royal Society was the official forum for debate, and the place where new ideas could be published and circulated quickly. For informal contact the coffee houses played an important part (Ellis 1956). Even Flamsteed, who never liked making the cold, damp journey up the Thames to the City, was often to be found at Toothes, Jonathon's or Child's, especially in his younger days. Hooke's diaries show that he was at the coffee houses two or three times a week. He met regularly with Halley, Moore, Gale, Wren and others in the 1670s. He frequently played chess with Theodore Haak, whose main interests were in theology and the arts, so that social contacts were reflecting and broadening his interests. Hooke mentions going to the theatre with some friends, but it was on the way home from Jonathon's coffee house that, for instance, Halley asked him 'whether one might use the other eye to look at the species of the sun while looking through a tube on the horizon'. Hooke's reply was predictable: 'I told him I had neer twenty years since shown the Society such an Instrument to tri it with.'

Hooke was extremely active and mobile. For the years of his diary, dining at home was a noteworthy event. Most of the time he dined out with other scientists. These contacts must have helped to circulate ideas and stimulate new lines of investigation, but it is noticeable that those who achieved the most in astronomy spent more time alone than Hooke did, and this is especially true as they grew older.

Hooke is one of the few scientists of the period to articulate views on collaboration. He wrote in his *Method of Improving Natural Philosophy* that collaboration and generosity in collaborating are two important qualities for a scientist. Coming from Hooke, who must have initiated more priority disputes than any other scientist in his period, this is an interesting view. Collaboration was becoming important. Generally up to this time there had been one scientist and possibly assistants involved in any one scheme. Hooke must have been looking in vain for someone to work with who would not try to claim priority over the discoveries that Hooke considered to be his. Almost everyone seemed to have done that to him.

As it was, the existence of the Royal Society itself is some evidence of a sense of community among scientists, but examples of genuine collaboration are rare. In astronomy, Flamsteed and his correspondents provide an example of the beginnings of collaboration. There was some sense of a partnership to achieve common goals, although in most cases the partnership was skewed by the admiration of the correspondent for Flamsteed, and the consequent humility in talking to him.

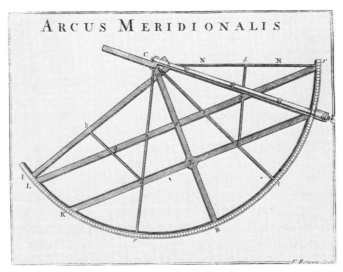

Two of Flamsteed's most important instruments, as described in chapter 7. *Top:* the 7-foot sextant set up ready for observing in the sextant house. *Bottom:* the meridional arc or quadrant. On the ceiling of the painted hall in the National Maritime Museum can be seen a painting by Sir Joseph Thornhill showing Flamsteed using the quadrant with the assistance of Thomas Weston. Royal Greenwich Observatory.

7

The tools of the trade

Telescopes in the seventeenth and early eighteenth centuries were used for a variety of functions, as they are today. They were used to inspect the celestial bodies in greater detail, to reveal features which could not be appreciated by the eye alone and, most frequently in England, to make measurements of the positions of the celestial bodies. In doing so, astronomers used telescopes either to take the time of phenomena such as occultations, eclipses or transits, or to measure the angles between bodies. Then as now, telescopes presented problems in optics and mechanical engineering as well as finance, politics and manpower.

Telescope optics may be constructed principally with lenses, in which case the telescopes are called refractors, or with mirrors, in which case they are reflecting telescopes, or reflectors. Throughout the seventeenth century refracting telescopes were in common use. This is the simplest form of telescope. Light passes down the tube to the eye through a configuration of lenses. The trouble with these telescopes was that the only way to improve the resolution with the simple lens designs then available was to reduce the curvature of the lens surfaces, in other words, to increase the focal length of the lenses; and this meant making the tube longer. As a result, attempts were being made to produce workable telescopes 200 feet long. Such instruments were as unhandy as the brontosaurus.

Lens telescopes also suffer from chromatic aberration, the rainbow-like coloured haloes seen around images even in modern lens systems of low quality. There is no chromatic aberration in mirrors. Exploiting this, Newton and Gregory in England and Cassegrain in France had developed by the 1670s, each in his own way, a type of reflecting telescope. In a reflecting telescope, the light from an object is admitted through an

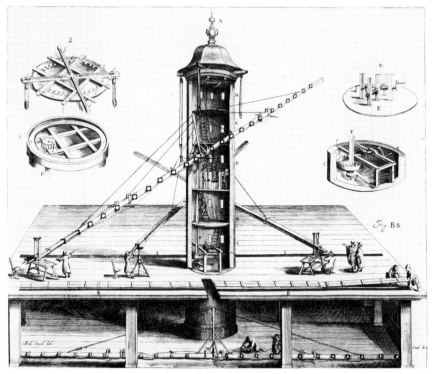

Figure 7.1 Long refracting telescope fixed to a tower. Illustration from Hevelius' *Machina Coelestis* (1673). Science Museum, London.

opening at the top of the telescope onto a concave mirror at the bottom of the tube. The mirror then reflects the light into an eyepiece. There are several optical arrangements which accomplish this. Reflectors not only offered much more satisfactory resolution than was possible with refractors, but also they offered the potential to be developed into better and better models for the future.

To produce these instruments there emerged a new group of craftsmen, the scientific instrument makers. Some of the older crafts had trained men who could work with intelligence and precision, and it was largely from these crafts that the instrument makers came. Many optical instrument makers came from the ranks of the spectacle makers and glass grinders. A large group of clockmakers became important in cooperating with scientists to make clocks and watches that would fill their particular needs. Among both the instrument makers and the clockmakers a few stand out because they could not only work to the most exacting hair's-breadth accuracy but also could innovate, could see new ways of solving problems.

In London could be found some of the best instrument makers in Europe.

John Yarwell, who worked at the sign of Archimedes and Spectacles near St Paul's, was himself a spectacle maker and member of the Spectacle Makers' Company. By the new century he had turned successfully to making telescopes and microscopes. His instruments were known and in demand on the Continent, as well as in England. Yarwell's chief rival was John Marshall, also a spectacle maker at St Paul's, who occupied the next door shop. Marshall emphasised in his advertising that his was the *second* shop within the gate on the left hand. He boasted that he was the only instrument maker whose work had the approval of the Royal Society.

Instrument makers could be involved in major developments in technology and were needed for the practical expression of new ideas. Richard Reeves and Christopher Cocks, one a glass grinder and the other a spectacle maker, worked together on polishing the mirrors for reflecting telescopes. Several scientists (Zucchi, Cavaliere, Mersenne and Descartes) had considered reflecting telescopes. James Gregory was the first to design and attempt to make a reflecting telescope in 1663. His design involved a primary paraboloidal concave mirror which reflected the light from a star to a small ellipsoidal concave mirror. The light was refocused by the concave mirror back through a hole in the primary mirror to an eyepiece behind.

The use of paraboloidal and ellipsoidal mirrors overcomes the problem of aberrations caused by spherical mirrors, but the figures of such mirrors are more difficult to make. Reeves and Cocks worked on the project of making various attempts to shape the new type of concave surfaces, but were unable to produce anything accurate enough. Gregory compromised his design and tried to construct a telescope with a spherical surface. The result was so unsatisfactory that he abandoned the project for the time being. Aspherical mirrors continued to give difficulties for some time to come. In 1697, Stephen Gray wrote to the Royal Society suggesting a method in which a sheet of pliable metal would be suspended over a ring until gravity made the centre subside into a perfect parabola. Gray had not tested his idea.

Newton's design for a reflector was less ambitious than Gregory's and, perhaps because of this and his great technical skill, he succeeded in actually making the first reflecting telescope. It used a single concave mirror and a small flat mirror at 45° to the optical axis which reflected the converging light to a hole in the side of the telescope tube and to an eyepiece. He used a spherical concave primary, blocking off the edges of the mirror and minimising spherical aberration.

Newton made the first of these telescopes in 1668, grinding the mirrors himself, but, characteristically, he made no effort to publicise his new design, and only a few close associates in Cambridge knew anything of its existence before 1671, when Oldenburg brought it to the notice of the Royal Society. Newton himself was stung to action in 1672 when news

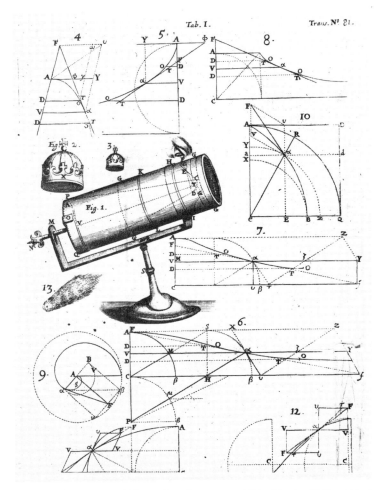

Figure 7.2 Diagram of Newton's reflector as published in the *Philosophical Transactions of the Royal Society* in 1672. Royal Greenwich Observatory.

came from the French Académie des Sciences that a Frenchman, M Cassegrain (his identity is not known in detail), had invented a reflecting telescope superior to Newton's. Cassegrain's design replaced Gregory's concave secondary mirror with a convex secondary. The combination of concave and convex mirrors reduces spherical aberration, as was later shown by Ramsden in 1779. However, Newton immediately issued a statement listing the disadvantages of Cassegrain's design, claiming that it contained no advance on Gregory's idea and, wrongly, that the system of mirrors would magnify the aberrations. At present, amateur astronomers usually make Newtonian reflectors, while professionals use Cassegrain's design; this illustrates that Cassegrain's design is superior but more

(4004)

An Accompt of a New Catadioptrical Telescope invented by Mr. Newton, Fellow of the R. Society, and Professor of the Mathematiques in the University of Cambridge.

THis Excellent Mathematician having given us, in the Transactions of *February* last, an account of the cause, which induced him to think upon *Reflecting* Telescopes, instead of *Refracting* ones, hath thereupon presented the Curious World with an *Essay* of what may be performed by such Telescopes; by which it is found, that Telescopical Tubes may be considerably shortned without prejudice to their magnifying effect.

This new instrument is composed of two Metallin *speculum's*, the one Concave, (instead of an Object-glass) the other Plain; and also of a small plano-convex Eye-Glass.

By *Figure* I. of *Tab.* I. the structure of it may be easily imagined; viz. *That* the Tube of this Telescope is open at the end which respects the object; *that* the other end is close, where the said Concave is laid, and *that* near the open end there is a flat oval *speculum*, made as small as may be, the less to obstruct the entrance of the rays of Light, and inclined towards the upper part of the Tube, where is a little hole furnish's with the said Eye-glass. So that the rays coming from the object, do first fall on the Concave placed at the bottome of the Tube; and are thence reflected toward the other end of it, where they meet with the flat speculum, obliquity posited, by the reflection of which they are directed to the little plano-convex Glass, and so to the spectators Eye, who looking downwards sees the Object, which the Telescope is turned to.

To understand this more distinctly and fully, the Reader may please to look upon the said *Figure*, in which

AB is the Concave *speculum*, of which the *radius* or semi-diameter is $12\frac{2}{3}$ or 13 inches.

CD another metalline *speculum*, whose surface is flat, and the circumference oval.

GD

Figure 7.3 Description of the working of Newton's telescope printed by Henry Oldenburg in the *Philosophical Transactions* of 1672. Royal Greenwich Observatory.

difficult to make. The Royal Society in 1672 lined up patriotic feeling behind Newton, and did such a good job of publicising Newton's priority that he was known for a while on the Continent as the 'telescope maker of England'.

In the early eighteenth century English astronomers worked with Newton's design as a result of his status and the weight of his opinion. In 1721 John Hadley produced a Newtonian telescope for which he had ground the mirror with the help of his brothers. Bradley and Pound tested it at Wanstead and found that it was much more convenient than the 123-foot refractor made by Huygens that they had been using. Reflectors were already beginning to look like the instruments of the future.

Up to about 1720, however, refractors were still the only telescopes in general use. A letter written by Hooke (Rigaud 1965) to give information to Hevelius in Poland tells a good deal about the best telescopes available in 1670. For not less than £25 Hevelius could acquire a 68-foot telescope (a refractor) with a single convex lens. The aperture would be adjustable between 3 and 4 inches. The tube would be made of very thin, light, slit deal bound at intervals along its length with thin iron plates. The whole would be slung on a handle like a pair of scales so that it could be raised or lowered as required. Hooke says that the whole tube of such a telescope does not weigh above 200 lb and, somewhat improbably, claims that ' 'tis manageable with the greatest facility imaginable'.

With a telescope of this size, Hooke was able to see many things not visible with a 36-foot glass, 'even a very good one'. He could see 'the shadow of the satellites and the verticity of Jupiter and Mars on their axes'. In other words, he could see enough detail to judge the orientation of the planets. In the next fifty years, Flamsteed was settling into the position of Astronomer Royal, and all his assistants and many correspondents were requiring telescopes good enough to see the occultations of Jupiter's satellites. Few of them had telescopes as good as the one Hooke describes. The most common was a 16-foot long refractor just adequate for the purpose. Stephen Gray, for example, had two telescopes and a 2-foot quadrant. At the very end of his life, in 1736, he was enquiring about the cost of a new 3-foot quadrant. He was told that it would have to be covered in brass to keep the termites out, and would cost £7. He died 19 days later.

Gray made no apparent attempt to acquire a reflecting telescope. He did all his work with two refractors of 6 feet and 16 feet respectively. In addition to these, he apparently possessed a theodolite. He described an occasion on which he saw a parhelion, or mock sun, 'not much unlike the sun when seen through clouds with its periphery not exactly defined'. When he saw this, he immediately rushed out to the garden 'taking a theodolite with me in order to take its distance from the sun'. This is the more surprising if you remember that Gray was a dyer by trade, and had no reason to keep a theodolite on hand.

Gray was ingenious enough and curious enough to develop his own methods for making glasses of the best configuration for telescopes. His letters on the best way of making a concave mirror show his interest and his ability. He successfully ground lenses for microscopes, as we know from his offer to make some for Margaret Flamsteed. In this field he was led by his own thoughts to explore the possibility of new experiments: 'Those Congruous Propertys known to be in small Drops of Water viz Transparence, Refrection and Sphaericity lead me to conjecture that they might if aptly Disposed be not unfit for Microscopes since they have the Requesite above mentioned that make the Glass Globules accelent ones' (26 May 1697). To investigate this idea further, Gray made himself a piece of brass with a hole at one end. Through this he was able to experiment with the magnifying effects of water droplets which, he found, would first of all form a plano-convex lens and then, if moved with a pin, would form a double convex lens. By moving this to and fro over an object he could see it 'little less distinctly then by glass Microscopes Especially by Candle which I finde much better then Day Light'. The Royal Society found this work interesting, and the letter was read aloud at a meeting and later published.

Gray's curiosity had led him to a subject of much contemporary interest. He was encouraged to the extent of two more letters describing a water microscope ingeniously made. Evidently his skills included working with brass and glass. He made use of many substances including quicksilver and isinglass. To polish his lenses he used a hare's foot. Alone in his dyer's workshop he achieved a great deal but, inevitably, he came across some technical problems: 'I have begun some experiments towards a way of making a large concave Speculum for Burning Glasses and have proceeded so far as to find Materials that will naturally receive their true figure though of many feet in diameter but have not yet overcome the difficulties of giving them a good polish' (12 May 1697).

As Gray had so many practical skills, he may well have made parts of his own telescopes. This seems likely, since the cost of a telescope was a minimum of about £25 and Gray was constantly forced to make compromises because he lacked the money to buy the books and instruments that he would have liked. In spite of the cost, most of Flamsteed's assistants and correspondents making an important contribution to astronomy had acquired a 16-foot telescope. Pound, for example, had a 16-foot telescope in Chusan for his survey of the southern sky. He set up the telescope on the roof of the house and had enough success with it to be able to observe some occultations of Jupiter's satellites. Unfortunately, Chusan is too far north for the sky survey to have been of great use to Flamsteed. In any case, Pound's observations were all destroyed in the native mutiny (see chapter 5).

When he returned to England, however, Pound lost no time in setting up another observatory in his house at Wanstead, where he was appointed rector. He was a man with a creative approach to technical problems. When he acquired one of the aerial telescopes invented by Christian Huygens he made use of a maypole to obtain the necessary height for the mounting. The aerial telescope had a focal length of 123 feet, but it was made without a tube. The object glass was at the top of the pole and the eyepiece was mounted on two wooden legs at the end of a thread or string.

The difficulties of observing with such a telescope were vividly described by Joseph Crosthwait, who went to visit Pound in 1720 in the hope of seeing the satellites of Saturn. When he got there, Pound told him that the sky was too light for Saturn's satellites but, instead, showed him Jupiter, which he saw clearly enough to be convinced that the lens was good. A brief look was one thing but a careful observation quite another. Crosthwait pointed out that the action of the air, the shaking of the pole, etc, 'renders it very difficult to trace the object and makes me conclude that not many good observations can be made with a glass of 123 feet in the open air' (6 May 1720). The reasonable tone of this comment is followed by a rather snide remark: 'However it had in some measure answered his designs, it having been the only means by which he and his kinsmen have obtained two good livings.'

Even to a more objective observer, Pound's telescope must have seemed difficult to use. In order to manipulate its great length, Pound set up a system of wires attached to the object glass and the eyeglass, at each end. Between these ran a narrow cord, with which the eyeglass could be turned by hand. By pulling the cord an observer could hope to adjust the object glass at the top. But, as Crosthwait pointed out, the cord varied in length according to the weather and Pound himself was forced to admit that 'this alone rendered the finding an object very difficult. But tis still more difficult to keep the object in the glass when you have found it.' Another hazard was the uneven ground round the maypole, which meant that an observer was likely to lose his balance and his object frequently in the course of a night. In spite of the difficulties, the glass was, as Crosthwait conceded, a good one. Wanstead was an important site for positional astronomy as James Bradley, Pound's nephew, installed a zenith sector there, and made important observations in the 1720s.

William Derham later took over the 123-foot telescope, but found it too difficult to use. The cost of remounting it satisfactorily would have been £80–90. He did most of his observing with a 16-foot tube instead. He gives few details, but, apparently, it was capable of being used to observe the jovian satellites when Derham could get around to it.

In his correspondence with Flamsteed Derham complained, gently but frequently, about the difficulties of practising astronomy, especially for a married man. His intention was to make regular observations of the jovian satellites, as well as to observe lunar and solar eclipses. He frequently wrote to Flamsteed, however, to report a fruitless year caused not only by bad weather but by too much company, 'busynesse' and, not uncommonly, 'forgetfulness'. Moreover, he declined to observe between 11 PM and 4 AM. Getting out of bed at midnight caused 'too much disturbance of my rest'. Not only was Derham unwilling to rise at unsocial hours on his own account, but, as he was one of the few married astronomers, he also had problems with his wife. Just how much she resented her husband's activities is not clear, but she was evidently not encouraging. In a letter of 27 December 1709 Derham mentioned 'the uneasyness of my bedfellow' as a sound reason for staying exactly where he was when a clear night might have beckoned a more single-minded man out of the warmth.

Thornton and Wright both had 16-foot telescopes, but Thomas Brattle in New England had only a small telescope of 4½ feet. With this he seems to have been mainly interested in observing lunar eclipses. In his letters he mentions his telescope briefly, but seems much more interested in his clock. Keeping clocks accurate was a problem for all observers, but especially for Brattle, who had to establish his exact position before he could begin. He tried asking Flamsteed for a method to establish the difference in meridians between London and Boston. In the end, he found a method for himself. He compared his own lunar eclipse data with those of Hodgson at Greenwich and, assuming that both were accurate, he was satisfied that he had the difference in meridians correct.

To convince Flamsteed that his timekeeping was accurate, Brattle described his clock. He had an eight-day clock 'with springs' that was so accurate that it did not lose so much as a minute. He checked the clock, comparing it with the timekeeping of three other instruments that he had with him. One of these was a ring dial. He also had an astrolabe and a 'curious brass quadrant'. He had evidently understood the importance of taking the refraction of the light from a star by the Earth's atmosphere into account in correcting his data. Flamsteed might have emphasised it to him as, in his letter of 1707, Brattle mentioned that he had checked the clock and found it 'a minute too slow without allowing for refraction'. As far as we know, Brattle used his quadrant only to check the time of his clock.

The quadrant was, however, one of the group of astronomical instruments most frequently used by contemporary astronomers for measuring the positions of the stars. In pre-telescopic times the instruments consisted of sights, as in a modern rifle, over which the astronomer looked at a star. The sights were attached to scales so that the angle could be measured, say, between the star and the vertical (defined by a plumb line).

When telescopes were attached to quadrants as sights there arose the opportunity for greater accuracy, since the star could be viewed with better resolution (up to 100 times more finely than with the eye). At the same time, the instrument maker was challenged by more demanding problems of engineering: the scales had to be made finer and more accurate. In part this could be achieved by increasing the size of the instruments so that a given accuracy in their construction achieved a greater accuracy in angular resolution. Indeed, it was necessary to make large instruments in order to mount the long refractors on them. But the additional weight of the larger instruments caused them to flex more, a problem more readily revealed by their potentially increased accuracy. Finally, instrument makers were challenged in making mechanical parts—screws and scales—to finer tolerances over the larger distances involved. This required geometrical knowledge in setting up the subdivisions and patient skill in creating them. There were many practical difficulties—Derham wrote to Flamsteed to ask how he could grind the limb of a quadrant too large to fit in a lathe, for example.

Quadrants were instruments for measuring star positions relative to the vertical: they could be fixed in a north–south wall so that the passage of stars across a known longitude could be measured. A small quadrant for the Royal Society made the name of the instrument maker, Thomas Tompion, even though his first quadrants were far from satisfactory. He gradually became known to a wider clientele until he was one of the best known instrument makers in London. In 1674, the Royal Society had decided to acquire a quadrant for its corporate use. Richard Shortgrave, a surveyor and instrument maker, was asked to make one, but was encountering the usual difficulties of grinding and adjusting. Not satisfied with Shortgrave's work, Hooke took the commission to a new man who had just made a quadrant for him and was setting up business in Water Lane. In rather an off hand manner Hooke referred to 'Tompkins of water Lane' during the first part of their relationship. It was not until they had met four times that Hooke felt sufficiently friendly to bother to get the name right, and to record that he met Tompion and they 'discoursed of founding, shrinking and swelling of metall, balls, sirens etc'.

By this time, Hooke was beginning to appreciate Tompion's ability. The quadrant for the Royal Society was finished in a few months, which, alone, was a recommendation, as some instrument makers would take years to finish a commission. The price was reasonable too at 'not more than £10'. Hooke introduced the quadrant as having been made by 'Mr Tompion . . . this person I recommend as having employ'd him to make that which I have, whereby he hath seen and experienced the difficulties that do occur therein' (Symonds 1951).

Although Tompion had made an attempt at a quadrant that Hooke thought admirable, for the time being, he had certainly not overcome all the problems. In 1678, Jonas Moore got hold of the Royal Society quadrant to lend to Flamsteed. Flamsteed, newly established at Greenwich, had only a small quadrant of 3 feet radius that he had brought with him from Derby. In his memoirs (Baily 1935), he commented that 'it was no very good contrivance but with it however I could take the sun's or a star's height so exactly that the difference betwixt the errors of the clock collected from 4, 5 and sometimes 6 several heights of the sun or a star was scarcely more than 10″ .'

The quadrant borrowed from the Royal Society was, Flamsteed grudgingly admitted, 'neater' but 'it was so ill contrived . . . that I could not make it perform better than my first'. This may have been partly sour grapes, as personal animosity towards Hooke played some part in Flamsteed's reaction. He says that he kept the Royal Society's quadrant until 1679 'when the ill nature of Mr Hooke forced it out of my hands'. If it were of such poor quality, it is surprising that it *had* to be forced out of his hands.

Flamsteed's needs for accurate and, therefore, large instruments were so urgent that he had to try his hand at supervising their construction. While he was still staying with Jonas Moore, he had commissioned the making of a sextant. This was an instrument carrying two telescopes which could be pointed at two stars in order to measure the distance between them. The frame was made in wood, but the axles and semicircles were to be cast in iron by the smiths at the Tower at Sir Jonas' expense. Unfortunately the chief workman died before the instrument was finished, and it was then that Flamsteed ran into serious difficulties: 'I could not make his successor to understand how the bigger semi-circle was to be moved by a perpetual screw; but forced to suffer him to move it by wheel work.' By the time the smaller semicircle used to incline the plane of the instrument was to be fixed the workman had understood the principle and 'wrought it off so that it performed very well'.

The sextant turned out reasonably well, and Flamsteed expressed himself satisfied with it (although he later made better ones for his friends). Yet he ran into that difficulty common to so many scientific enterprises: overspending. The workmen charged so much for the sextant that Moore took fright and 'Mr Hooke persuaded him to leave the contrivance of a ten foot quadrant to him. Upon his promising to frame it at a lesser charge. I was not to see it till finished.'

As Flamsteed expected and maybe secretly hoped, Hooke's quadrant was a disaster. To avoid making an accurate fine scale of ten foot radius

Hooke subdivided the scale coarsely, and provided an 'index', which was locked at the coarse subdivisions. It supported the sights which could be moved on a fine five-degree scale relative to the coarser subdivisions. The weight of the index made the five-degree scale flex so that the quadrant could never be considered reliable. Moreover, the limb was two inches short of a true quadrant. From Flamsteed's point of view the instrument was quite useless. He made clear to Jonas Moore that he still desperately needed a good wall quadrant, but 'still all my complaints were insignificant in those ticklish times'. The only remaining possibility was to try to contrive some means of using the sextant to take meridional heights. By December 1676, he had begun to work with the sextant: 'I soon found a way to fix it in the meridian; and by a particular contrivance, whether it altered its position whilst I moved the index.'

Using this contrivance, whatever it was, to keep some check on his accuracy, Flamsteed seems to have satisfied himself of the sextant's reliability during its manufacture. However, when he tested the instrument again by checking its measurements against the greatest and least heights of some stars whose positions he knew, he found the error of the instrument to be considerable. Later, he found that there was another unpredictable error caused by movement of the limb of the instrument itself. However careful he was in working with it, the sextant was liable to flex to the extent that, in taking the height of the same star on different occasions, his results would vary by as much as half a minute and sometimes more.

Flamsteed's care and integrity as an observer are revealed by his attitude to these observations: 'and though I made some number of observations of meridional heights I inserted none of them into my catalogues. I knew they ought to be better determined and I hate to recommend anything to the public of which I am not very certain.' Writing this in retrospect, Flamsteed then paused to consider the effect of inaccurate observations: 'Coarse observations made by honest, well-meaning men have more perplexed the astronomer than all their labors and dreams upon them can make him satisfaction for.' Some observers may use inaccurate observations to serve their present turn and may appear to support their 'pretty thoughts and conceits' but, in the end, they will be found to be what they really are: mere theories with no solid foundation, and then they will 'turn to his shame and reproach', as did the work of Lansberg and Riccioli.

Flamsteed had no intention of following in their footsteps. He would present to the public nothing he could not proudly assert to be as accurate as he could make it. As long as he had nothing but the sextant, he put his mind to making the best possible use of it. He devised a method of measuring the Earth's position among the fixed stars and the 'inequalities' of her motion. If he took the Sun's distances from either Jupiter or Venus in the day time, he could then take the distances of one or other of these from various fixed stars at night. This gave him the position of the Sun relative to the stars, which is a reflection of the Earth's position. Its departures from simple circular motion are the inequalities referred to.

This enterprise was all the more attractive because there was a difference of 8 minutes of arc between the equation of the Earth's orbit as deduced from Tycho's and Cassini's observations and the equation that Kepler deduced. Flamsteed was anxious to produce tables that would settle the controversy once and for all. Using Tycho's longitude for the First Point of Aries, he produced a corrected figure for 1679 and set to work. He finished the tables that same spring. Nevertheless, he admitted to some misgivings; having just made clear his abhorrence of inaccurate observations his own honesty made him point out: 'And though the sextant was not so steady an instrument as is requisite for this purpose, yet it was much better to depend upon it than have conjectures till such time as I could procure a fixed wall quadrant for this purpose.'

In spite of the difficulties involved, Flamsteed saw the making of these tables of the Earth's positions relative to the Sun as the first step towards his eventual publication of a catalogue of fixed stars that would be more accurate than those of Tycho or Hevelius, and very importantly would 'need no correction by those that come after us'. He pointed out to Jonas Moore, while he was working on these tables, that the whole task took Tycho 20 years and Hevelius 30 years. Moreover, he added significantly, they were both far better provided with instruments and assistants than he was. In fact, Flamsteed was aware that telescopes on the Continent were generally better than those available in England. He wrote gloomily to Richard Towneley that he had been testing long telescopes and found them poor: 'I have got the glass of 60 ft tried but find neither it nor Dr Coxe's of 48 ft to be any ways excellent. They show Jupiter large enough but I think I saw his belts distincter in a 16 ft glass than in either. I am confident by those observations that Mr Casigny lately imparted that the French glasses excel ours far. How we shall get any better I know not.'

One thing that the English could do better than the Europeans, or so Richard Towneley boasted to the Council of the Royal Society in March 1667, was measuring the minute portions of seconds necessary for astronomical observations of close stars. An English astronomer from Lancashire who died in battle during the Civil War, William Gascoigne, had invented a device called a micrometer. Flamsteed remarked that the micrometer was a 'curious though little engine applied to a telescope' (16 January 1674), and he wrote to W Molyneux on 10 May 1690 that the first valuable measurements that he ever made were taken when he was given a micrometer. In this little engine, two pointers, either the straight edges of two strips of metal or two hairs, were positioned in the focal plane of a telescope lens and mounted on a threaded screw of fine pitch. The screw was rotated with a graduated scale, and the number of subrevolutions of the screw measured the amount by which the two pointers were closed or opened. The screw and scale had the effect of mechanically magnifying small motions of the markers. With this device, positions and distances could be measured as accurately as the telescope would allow.

Towneley made various improved versions of Gascoigne's micrometer. His first was capable of measuring 1307 parts to the inch. Later ones showed over 3000 parts to the inch. A micrometer made by Towneley was presented to Flamsteed at Jonas Moore's house and was a valued part of his equipment. Nevertheless, it could never do more than measure what the optical instrument showed. In order to compete with the Europeans and produce his own tables, Flamsteed always came back to the need for new instruments.

Jonas Moore died in August 1679, leaving Flamsteed even more anxious about the payment of his salary and with very little hope of any allowances for more instruments. The only course left to Flamsteed was the same as that adopted by many of his country gentlemen. He would have to make his own instruments. His first attempt was far from satisfactory. In August 1681 he began work on a small quadrant which he designed to be of the same radius as the sextant, so that it would show the meridional distances of stars as well as the sextant showed their distances from each other.

As in making the sextant, the main problem turned out to be the labourer. There was no pool of skilled labour for this kind of work, as such skilled craftsmen as there were had plenty of employment in the London shops. Since Flamsteed was trying to keep costs down, he was using his own servant, trying to teach him the necessary skills to be helpful. The servant was probably Cuthbert Denton, and Flamsteed had continual problems with Cuthbert, who went to town and never returned before midnight. He complained that Cuthbert was an 'ill workman who respected nothing but the getting of wages by his work'. The result was that the arm of the quadrant was not well made and for two years he abandoned the project in disgust. Then he took it up again, as it still represented his only hope. He fixed the quadrant to the wall in the meridian and divided the arm into degrees beyond the position of the pole star. This was a measure to make sure that the instrument would remain useful even if its position should accidentally alter. It would no longer give meridional heights accurately, but could be corrected by observing star transits above and below the pole. It could then still be used to show accurate distances of other stars from the pole. A shift was exactly what did happen and this foresight was well rewarded because, although the quadrant could not be used for the full extent of its intended function, it could still give some useful observations.

Although this quadrant had limited use and was not the instrument that Flamsteed had hoped it would be, it gave him valuable experience and showed him the importance of some of its features. He had made the arm show an arc of 140° in the sky and this, he was convinced, was a most valuable innovation, as it allowed him to see all the stars visible at his latitude. This he decided to incorporate in a good quadrant whenever he might get the chance to make one.

The chance came when Flamsteed's personal income was greatly increased by inheriting his father's estate. He records 'my good aged father dying . . . I resolved to employ some part of my estate (which was increased by what he left me) in building such a strong mural arc as I had long before designed' (Baily 1835). When he told Lord Dartmouth, who had replaced Jonas Moore as Master of the Ordnance, what his plans were, Dartmouth promised to reimburse his expenses. In fact, he never did, but Flamsteed went ahead with a good heart.

Other circumstances contributed to the success of this quadrant. Cuthbert Denton had gone and another, rather better, servant, John Stafford, had died at the Observatory in 1688; Flamsteed was free to employ someone new. He had already seen the work of Abraham Sharp and seized on the opportunity to hire him at once. Sharp was 'not only an excellent geometrician but (which I no less valued him for at this time) a most expert and curious mechanic.' Sharp and Flamsteed together spent three months on making the frame and attaching it to the wall. Flamsteed invented a new method of planing it once it was in position on the wall so that it would lie flat and not cause any inaccuracy by being in any way bent or uneven.

The next stage took almost a year to complete. The divisions had to be calculated, accurately marked and engraved. Here the difference between working with Denton and working with Sharp became most clearly apparent. Sharp worked with constant care and vigilance to get the thing right. An aspiring astronomer himself, he understood the purpose of the instrument, understood the need for accuracy, and was interested in how it could be made.

When the engraving of two sets of divisions and figures was finished, the instrument was carefully tested. Results from observing distances with the sextant were compared with results from observing transits with the quadrant. The two sets of figures were so very nearly the same that Flamsteed felt able to conclude that both methods of observing were equally good, but that the quadrant was much easier to use and better for his health, although he never says exactly why this was so.

Before he could settle down to do his main life's work with the quadrant, however, Flamsteed had to face one more problem. As soon as the quadrant was finished, in November 1689, he made some observations of distances of southern stars from the vertex. Later, when he had rectified the instrument, he checked these distances and found them as much as 60–65 seconds of arc bigger than they ought to be. At the time, he concluded that the instrument had been slightly bent while he was working on it. Later, however, in December 1690, he was taking the position of the pole star, and found to his surprise that its distances from the vertex had increased and, on the other hand, the stars passing the meridian in the south were decreased by the same amount. The explanation then occurred to him: the northern end of the wall to which the quadrant was attached had subsided. Once this was established, the proper correction could be made.

Flamsteed describes his observing technique in his memoirs, so that we can picture him at work. He needed only one assistant for observing with the quadrant. The assistant's job was to count the clock and write down, to Flamsteed's dictation, the raw data. With the sextant, however, because of the greater difficulty of manipulating it and the two separate telescopes, two assistants were scarcely enough. Because of the convenience and accuracy of the quadrant, Flamsteed had no desire to use the sextant again.

The quadrant was an improvement over the sextant also in the way it was set up. A problem with the sextant was that an observer looked towards the open sky, attempting to focus on a particular star. If a bright star lay nearby, he was unable to see a star of sixth magnitude or fainter at all. For the quadrant, Flamsteed constructed a covering with a slit that was only one and a half feet across. Through this slit, he found that he could easily see stars of seventh magnitude or less with his naked eye, and could conveniently make observations and note them down with his other main observations. Because he did so, his observing notebooks offer fascinating clues for modern astronomers to the appearance of phenomena at the time in which Flamsteed himself did not see any significance. He observed a star that may in fact have been Neptune and a small star in Cassiopeia that he called 3 Cass. This last may have been the object that has recently been identified as the supernova remnant known as Cassiopeia A.

One more advantage of the quadrant was that Flamsteed had it fitted with telescopic sights. These, alone, would enable him to make observations that would be superior to Tycho's made with plain sights. 'The collimation [alignment] through the telescopical index was perfect, easy and accurate; and therefore the observed distances from the vertex exact which his, certainly in the moon, were not'. Because of his telescopic sights, Flamsteed felt confident that he could be more exact in timing transits than Tycho could have been and would, therefore, be able to give right ascensions more accurately. This was not obvious to everyone, however, and Hevelius insisted that he could work as well if not better with plain sights than with telescopic sights. Halley was sent by the Royal

Society to evaluate Hevelius' claim in 1679 and came back with favourable reports, saying that Hevelius was getting results that were accurate to within five seconds of arc. Hevelius was an exceptionally careful and painstaking observer with great skill. Moreover, he must have had exceptionally good eyesight. After him it is not likely that anyone else could have maintained much of an argument in favour of plain sights. As telescopic glasses were constantly being improved, the gap between what could be done with the naked eye and what with glasses was bound to be continually widening. Even as far back as the 1670s the majority opinion favoured telescopic sights, and Flamsteed enjoyed the support of Hooke in this opinion.

Apart from his quadrant, Flamsteed's other essential instruments were his clocks. He had pendulum clocks made by Tompion, who had also helped in the delicate business of screwing the index to the limb of Flamsteed's sextant. Flamsteed's clocks had pendulums of 13 feet. When he described them in his memoirs of 1675–83 he sounded well satisfied with their efficiency. Each vibration of the pendulum took two seconds and the weights needed to be lifted only once in twelve months. Their timekeeping agreed with the times he arrived at by his Derby quadrant, usually to less than ten seconds.

These clocks had been given to Flamsteed by Moore when the observatory opened. They needed some adjustment when they were first put in place. He wrote to Moore: 'I am not much solicitous about our clocks since I doubt not but Mr Tompion's dexterity will put them soon into such good order as that a little pains of mine in some weeks may get them into good going again' (6 November 1677). Apparently, a small alteration in the works had caused one clock to go eleven minutes a day too fast. Flamsteed wanted to get them going satisfactorily and then leave them without further alteration. Unfortunately, Tompion was more interested in creation than in maintenance. Flamsteed found it difficult to lure Tompion down to the country to work on the clocks. A month later he wrote to Moore again: 'One of our clocks goes well, the other may be made to do so too if Mr Tompion could be prevailed with to come and bestow a little pains upon it' (17 January 1678).

Tompion was a busy man. A letter from Newton to Flamsteed (26 January 1695) mentions 'two of Mr Tompion's tables of the equation of time'. Apparently Tompion's tables differed somewhat from those of Moore and Flamsteed. The differences were small, '4″ or 5″ only', and show that Tompion had been at least competent in constructing his tables. Mixing with the friends of Hooke frequenting the London coffee houses had enabled Tompion to participate to some extent in the sciences for which he made instruments.

Flamsteed's instruments are an interesting, if small, sample of the tools of astronomy at the time. Even the few that he had were the subject of various disputes. When Jonas Moore died his son, young Jonas Moore, who Flamsteed had said would 'never be old Sir Jonas', demanded that the sextant and two clocks in Flamsteed's possession should be returned to him as part of his patrimony. Flamsteed, however, was able to prove to the Board of Ordnance's satisfaction that Sir Jonas had given them to him entirely.

After Flamsteed's death, a more protracted dispute arose between the Board of Ordnance and Margaret Flamsteed. Flamsteed had died at the end of December 1719. By March 1720, Halley had been appointed Astronomer Royal in his place, and was insisting that Mrs Flamsteed should leave immediately. With the help of Joseph Crosthwait, who stayed on most loyally in spite of the fact that Flamsteed had left him nothing in his will, she managed to gather up the papers and books and remove all the instruments. On 7 March she left, and on the same day Halley moved in.

When Halley found that all the instruments had gone and that he was in the same position as Flamsteed when he was first appointed, or perhaps worse in that he had no immediate provision from the Board of Ordnance, he set about acquiring instruments as best he could. The Board's reaction was to revive the claim for the sextant, Tompion's clocks and the books that Moore had given. The claim stated that these things had been given to the house, not to the man, and should, therefore, be left for Halley.

Flamsteed's original defence against young Jonas Moore was produced. Crosthwait thought that it would not be sufficient 'since they have the Crown to dispute with' (17 March 1720). The course of events can be traced in the correspondence between Crosthwait and Sharp as they worked on getting the catalogues printed as Flamsteed would have wished. Crosthwait gave Sharp all the news of London, as he was still living far away in Yorkshire. Halley, as might be expected, was faced with official inertia. In July he borrowed a quadrant from the Royal Society, possibly the one that Flamsteed had found so unsatisfactory. Halley had the same trouble and Crosthwait comments 'but tis so ill made that he cannot use it' (16 July 1720). Faced with the total inadequacy of this quadrant Halley was prepared to compromise: 'he would now quit his claim to the two clocks and everything else if Mrs Flamsteed would part from the sextant.' Mrs Flamsteed had no intention of parting with the sextant except to sell it, and she sent a message to Sharp asking that he would make no comment if asked what he knew of it.

In August, the position had not changed: 'Dr Halley has not yet got any instruments besides the quadrant I formerly mentioned, and I am now in more hopes than ever that he will not be able to get the sextant from Mrs Flamsteed' (20 August 1720). There had been a threat of a law suit but no sign yet of one being actually brought. In October the Office of the

Ordnance was becoming tired of Halley's insistence and seemed anxious 'to get rid of him for I am confident he has no friends amongst them at this time'. To get the affair settled, the Office was ready to abandon all claims except for the sextant and even to 'allow Mrs Flamsteed something for it' (8 October 1720). This was still not the full ownership that Mrs Flamsteed was holding out for, and the dispute dragged on.

In December, the rumour was that the Attorney General had been consulted and had given his opinion against the Board because they could produce no evidence of having made or repaired any instruments at all at their own expense. Consequently, if Mrs Flamsteed could prove that her husband had ever repaired the sextant, that would be a strong indication of her right to it. By April nothing had happened, except that Mr Molyneux was considering buying the mural arc and a quadrant.

By June there was still no sign of action. Halley had no thought of buying the mural arc himself and, in fact, had pulled down part of the meridional wall to which it had been fixed. To Crosthwait's great interest, he 'built a little boarded shed between the study and the summer house and has fixed a stone in the ground which stands about four feet high'. This was an object arousing some speculation: 'what he intends to fix upon it I cannot yet learn but as yet he has done nothing.' Crosthwait is interested to note that Halley still has not hired an assistant but 'he bears such a very bad character that I believe he may make observations by himself' (1 June 1721). If Crosthwait means by this that only a man of bad character would be sufficiently irresponsible to make observations without an assistant, this would be an interesting comment on the standard of practice instilled by Flamsteed in his assistants.

Halley's problems dragged on for some time. By December 1721, no-one had bought any of Flamsteed's instruments and Halley, in Crosthwait's opinion, had shown his essentially dilettante and trivial approach: 'Mr Molineux has not purchased the instruments and Dr Halley has converted the sextant and quadrant houses into a pigeon house.' Crosthwait saw Halley's persistence in refusing to buy Flamsteed's instruments or to use his installations as a deliberate condemnation of Flamsteed's methods and an attempt to deny any value in his results. Halley did decline to make use of Flamsteed's sunken wall but knowing how much it was at fault, he could have been accused of laziness if he had made use of it.

Halley's opportunity to set up instruments as he wanted them came in 1724, when the Board of Ordnance allocated to him the magnificent amount of £500 to be spent on instruments as he thought fit. In addition to this, he later received another £100 per annum salary through the Navy Office. Crosthwait speculated that he might at last buy Flamsteed's instruments as they would be cheaper than new ones and he might then 'put some of the money in his own pocket' (10 February 1724).

Halley was not as mercenary as Crosthwait thought and, in fact, had new instruments made to his own specifications. By July 1725 he had an 8-foot quadrant made and was already having it divided into measurements by the clock maker George Graham. He had Flamsteed's sunken brick wall removed and, instead, built a stone wall further away from the edge of the hill. In this he learnt from two of Flamsteed's mistakes. One was building in brick, which is likely to warp, and the other was in building too near the edge of the slope. Crosthwait's loyalty to Flamsteed can be measured in that he saw both of these changes as insults to Flamsteed and to the catalogue rather than as sensible improvements.

A few other types of astronomical instruments were in use at the time, though they met with varying degrees of success. Flamsteed had installed a telescope in a well but he used it only once in the summer of 1679. This tube was sunk down a well shaft. The intention was that the telescope could be held rigid in this position and would not suffer from the movement and flexing of long exposure in the open. An etching by Francis Place shows a spiral staircase down the well and a couch for the observer at the bottom. Flamsteed hoped to use it to measure very small variations in positions of stars. It turned out to be unusable because of the damp in the well shaft. As an alternative to the aerial telescope, the well telescope was an unqualified failure.

The Sun is of interest to astronomers, as well as the stars and planets. To observe its activities, special instruments and techniques are needed. From Stephen Gray came detailed descriptions of the techniques and problems of observing sunspots. He used both his telescopes, the 6-foot and the 16-foot, on occasions. He would focus his telescope so that its light was admitted on to a white card lying in the axis of the telescope. On the card were two lines at right angles to each other, one representing the parallel of the declination of the Sun's centre and the other the meridional circle, or the continuation of a north–south line on the Earth. With the 16-foot telescope Gray would observe and draw the shape of the spots, then, with the 6-foot telescope, he would observe their positions in relation to each other and the lines on the Sun. He had a good clock with a pendulum that vibrated every second. When the first limb of the Sun reached the north–south line on his card, he would then begin to count until the spot reached the same line. At this point he could write down the temporal distance of the spot from the Sun's limb. He also measured the spot's distance to the north or south of his horizontal line. All this he wrote down himself in his table of observations. In this way he was able to manage the whole business alone. On other occasions, Gray had difficulty in making all the observations and recording them by himself. When he observed a solar eclipse he had 'several spectators but no assistant' and probably the spectators got in his way because he regrets having 'lost the opertunety of seeing the degree the moon made her exit' (4 May 1706). Always ingenious, Gray described his own individual method of drawing a meridian line on the floor of his observing room so that he could orientate the Sun's image. He erected a brass plate with one small hole in it over a window in the roof. When the image of the Sun appeared on the floor in the morning, he made four marks on the floor to define the Sun's periphery or edges. He then found its centre. With a pair of compasses 6 feet long that he happened to have he was able to draw an east—west line. He put one foot immediately beneath the hole and measured the distance to the centre of his image. He then swung the compass round to draw a circular arc for the afternoon. When the Sun reached the afternoon point, he drew a mark on the floor and then joined his two images. In this way he obtained an exact east—west line. Perpendicular to this and passing through his central hole would be the north—south or meridian line. This he had done some years before but had checked it since and still found it accurate. He checked the time of his clock before the eclipse by observing the Sun crossing the meridian line. He also used it to hang plumb lines in the same plane so that he could see stars crossing the line and so measure their right ascension. These arrangements were made for the solar eclipse of 1706.

Gray was satisfied with his meridian line, but it was unlikely to have been accurate to closer than half a degree. Flamsteed pointed out in a letter to Richard Towneley as early as 1677 that the only accurate meridian line would be one drawn by the pole star rather than the Sun. You would need an open roof and, even then, there was difficulty because of lack of agreement on the exact right ascension and declination of the pole star. Even if this were certain, the refraction of the rays would create further inaccuracy. Then there are other problems like the unevenness of most floors and the inaccuracy of instruments for drawing straight lines on such a scale. Flamsteed himself had used the method, though, and, in fact, he set his meridian wall by the pole star. To find the error of his clocks, Flamsteed attached two 5-foot telescopes to a wall so that he could take the times of well observed stars crossing the meridian.

By 1733, the basic observing method for observing eclipses and sunspots had not changed at all, but Gray refers to his instrument and the card together as a 'helioscope': 'Our observations were made with an Helioscope consisting of a Telescope of 6 foot long fixed to a box 2 foot long at the end of which was placed the digit scheme the diameter of the sun's image being 6 inches' (19 May 1733). The time was kept by a clock checked by drawing a meridian line on the floor of the Great Hall in Otterden Place. A brass plate was attached to a window in the roof. Gray and Wheler checked the timekeeping by the meridian line both before and after the eclipse. In all of this there was no fundamental change in the method.

The major developments in astronomical instruments over the period were not reflected in the actual observing techniques of the majority of astronomers. The cumbersome sextant and quadrant were adequate for Flamsteed's own thorough, if unexciting, work, and astronomers of his generation were almost all still using refracting telescopes. There was plenty of scope for working with these in the areas of interest at the time. When the eighteenth century took off into galactic astronomy the new generations of reflecting telescopes were to come into their own.

8

The astronomer's image

Astronomy was an ancient branch of knowledge. The Greeks had assigned one of the nine muses, Urania, to be responsible for the divine inspiration of astronomers. In the seventeenth century astronomy detached itself from the philosophy of Aristotle and set out on the quest for truth based on measurement and quantity. At the same time astrology was defined as a different activity and was being hived off, no longer respectable for practising astronomers.

By the end of the seventeenth century these movements had become clear and the practice of astronomy was settling into its modern shape. For astronomers who lived then the general trends were apparent, but the degree to which they affected behaviour varied with individuals. All had to reach their own compromise between practice and theory. All had to decide what limits to put on the activity to be called astronomy. Where did astronomy end and the occult begin? For most of them, the Christian religion was an unquestioned assumption, but only a few saw their scientific work feeding directly into their religious beliefs.

In addition to the uncertainties generated by living during a major transition, astronomers, like all other scientists, had to endure the opinions of their contemporaries, with much criticism stated and implied. External criticism may be unpleasant, but it could be expected to give some sense of unity or group membership to those criticised.

One essential requirement for membership of the group by the second half of the seventeenth century was respect for observation and experiment, and a greater or lesser degree of scorn for philosophy unrelated to practice. In his quarrel with Hooke, Flamsteed hurled the name 'philosopher' at him as a term of abuse: 'but he proposed himself to be a Philosopher and not an Astronomer and therefore I have no more to say to him but that Astronomye is ill served when it must be ordered by the whymsies of Philosophy' (letter, 2 April 1678).

The impression that this accusation gives is unfair to Hooke, who was a brilliant and careful observer. It illustrates, however, another axis along which scientists might vary: the degree to which they specialised. Hooke was still a son of the Renaissance in the breadth of his scientific interests. Hooke's law was an advance in physics. His *Micrographia* was a scholarly work in microscopy and biology. Curator of experiments at the Royal Society, he found no subject outside his range and his interest. He designed instruments and worked on the casting, grinding and adjusting himself. After the Great Fire of 1666 Hooke became an architect and worked with Sir Christopher Wren (who had himself been no mean astronomer) on redesigning and rebuilding London.

Newton worked in astronomy from time to time but spent the years from 1696 wholeheartedly involved in organising the recoinage at the Mint until he was elected President of the Royal Society in 1703. In his early years, Newton's interests had been diverse. From about 1678 he was engrossed in chemical experiments. The detail of these experiments is difficult to understand because he used a private code of symbols in his descriptions and accounts, but his overall purposes seem to have been in the realm of alchemy. Thus, he tried to find correspondences in chemistry with the myths attached to the names of the planets. The extreme care with which Newton carried out his experiments, some of which took him years to complete, is one of the paradoxes of the seventeenth century to a modern mind. Scientific method applied to mediaeval alchemy seems incongruous. Yet he turned from his attempt to make sense of the universe in chemicals to his attempt to make sense of the solar system. To his mind it was all part of the same quest.

Halley, too, was far from devoting his whole time to astronomy. He spent much time and energy on other branches of science even when he was in London, working for the Royal Society. He was interested in all branches of physics, and he, too, like Newton, was happy to venture into cosmology. His careful studies of the rate of evaporation of water led to an interest in the Biblical Flood, one of the favourite puzzles of geologists and theologians alike. Halley approached it from the point of view of calculating the amount of rainfall necessary to cover the Earth. At the maximum rate of rainfall that he could envisage Halley could cover the Earth to a depth of only 22 fathoms, which would hardly have covered the lowest hills. Instead of postulating a miracle at this point, as the theologian Burnet and physician Woodward had done, he suggested that a comet passing close to the Earth could have changed the inclination of the polar axis. This became a popular hypothesis and was followed, notably by Whiston.

Halley also worked in areas that had no connection at all with astronomy. He studied tables of mortality and produced the first statistical tables of life expectancy. He made forays into classical scholarship, working out from a study of ancient texts and tide tables that Julius Caesar must have landed north of Dover at Deal or Sandwich. His experience at sea led to an interest in diving and the invention of a method for supplying fresh air to a diving bell by a system that involved raising and lowering sealed buckets filled with air. The same principles were to be applied to the raising and salvaging of wrecks and so had very practical applications.

Halley's experience on the East India Company ship that took him to St Helena emboldened him to take command of his own ship and set off on a voyage to the South Seas in 1698. The ship was called the *Paramour Pink*. Possibly because of his inexperience, Halley had trouble with his officers. At one point his lieutenant flatly refused to sail the course that was ordered. The lieutenant became so insubordinate that Halley was forced to return to England to replace him. The tension that must have arisen under the command of a landsman, whether or not he was a scientist, is suggested by Halley's account of the mutiny: 'because perhaps I have not the whole Sea Dictionary so perfect as he, has for a long time made it his business to represent me to the whole Shipps company as a person wholly unqualified for the command their lordships have given me' (Macpike 1937). Halley sailed the ship back from Newfoundland with the lieutenant confined to his cabin. In 1699 he set sail again without a lieutenant at all. His seamanship must have been adequate as he successfully completed his voyage and returned home.

Most of the other important astronomers of the period spent some time working in other sciences. Jonas Moore, who surveyed the fens, was also interested in ballistics. Earlier in the century, the breadth of experience and interest had been even greater. Wilkins, for example, had been interested in grammar and education, as well as science. His contemporary, Wallis, wrote a treatise on the grammar of the English language and even invented a language for the deaf. His interests extended to the fine arts and he wrote several books on the cultivation of taste.

Unlike Wilkins and Wallis, later astronomers began to limit themselves to science in its modern sense, leaving the arts to others. Pope, by the early eighteenth century, expressed the view in his *Essay on Man* that even science had become too complex for men to be widely versed in many fields:

> *One science only will one genius fit,*
> *So vast is art, so narrow human wit.*

The only seventeenth-century astronomer to confine himself almost exclusively to astronomy was John Flamsteed. Amateurs such as James Pound and assistants such as Abraham Sharp kept their interests directed to astronomy, but Flamsteed was exceptional in devoting his working life to it, with a very few ventures into related fields such as calculation of the tides.

Most amateurs were far from giving astronomy the whole of their time. The Royal Society's journal, the *Philosophical Transactions*, provided a forum for short papers, so that a piece of work on a fashionable subject of enquiry was likely to be published. There must have been a certain attraction, too, in the Society's avowed intention to ignore the social class of its contributors and consider only scientific merit. Robert Anderson, a silk weaver who knew his Virgil, wrote to the Royal Society to put forward some mathematical propositions, pointing out that sometimes '*de plebe* [among the common people] are found very intelligent and sagacious persons' and that he will therefore venture to contribute (*Philosophical Transactions* 39, September 1668). Not everyone was ready to talk to the lower classes, though, even in the interest of science. Evelyn found that he could not support 'conversing with mechanical persons'. Boyle, on the other hand, felt strongly that science should ignore social class: 'he deserves not the knowledge of nature that scorns to converse even with mean persons that have the opportunity to be very conversant with her.'

The majority of scientists would probably not have liked to have much contact with the lower classes. Spratt, in his *History of the Royal Society*, written in 1667, characterised the Society's membership at the time: 'But though the Society entertains very many men of particular professions; yet the farr greater number are Gentlemen free and unconfin'd.' The reason for this, according to Spratt, was that too many tradesmen in the Society were likely to mean too much interest in 'present profit' and, also, there would be inequality felt among the members: 'some imposing and all the others submitting and not as equal observers without dependence'.

In spite of the reservations, the Royal Society undoubtedly benefited when a silk weaver was able to publish alongside peers of the realm if he had something to say. Stephen Gray, the dyer, wrote confidently to the Secretary in 1700: 'Being well assured that the smallest Indeavours towards the advancement of Naturall Knowledge are allweis Exceptable to the Royal Society from whomsoever they proceed I Presume to submitt a few Experiments to their Consideration.' The subject matter, no doubt exactly the sort that would convince Swift that his satires in *Gulliver's Travels* were justified, was the rate at which sand would flow through different sized apertures. On the first glance, this seems a trivial exercise, but Gray's purpose was to try to develop a more reliable kind of hour glass that ships at sea would be able to use as a chronometer. The ultimate aim was, therefore, an improvement in navigation.

To an outsider, studying electricity must have seemed to be one of the most trivial games played by the scientists. Stephen Gray was fascinated by 'the luminous and electrick Effluvia' that Francis Hauksbee was producing for the Society by rubbing a glass tube. Gray had carried out a series of twelve experiments on the properties of static electricity, written them up and sent them to Sloane in January 1708. At this point the Society, particularly the *Transactions*, was under fire from outside for the triviality of its concerns. Gray wrote: 'Those that have any Delight in the knowledge of nature cannot but esteem them whatever some may say whoe had rather lay the Deficiencies they Pretend to finde in them on your Transactions then on their own unPhilosophical inclinations.'

Those with 'unPhilosophical inclinations' could see no point in studying these phenomena. Newton himself was inclined to think that time could better be spent on other studies. Besides, the effects of electricity were so little understood that they must have seemed to border on the occult. When Gray caused sparks to fly from the bodies of servant boys the general public may well have thought that the devil was involved. Certainly, the country people near Otterden Place, where he carried out experiments with his friend Granville Wheler, circulated rumours that the devil was being summoned on some nights to the big house. Wheler was known locally as 'the wizard'.

Stephen Gray was also involved in a manifestation of occult phenomena at Canterbury. On that occasion he was not in any way the cause. Instead, he performed a scholarly and very solemn investigation into a most bizarre ghost story. The apparition of the ghost of Mrs Veal to Mrs Bargrave has been made famous by Defoe, who wrote a short story based on the occurrence. In 1705, a 'very important person' heard of some supernatural manifestation at Canterbury and asked Dr Arbuthnot for more information. Arbuthnot asked Flamsteed, who asked Gray, who obligingly reported back. A woman had been visited by the ghost of a friend whom she did not yet know to be dead. Gray's conclusion was that the woman was a 'religious, discreet witty and well accomplished Gentlewoman'. Some aspersions had been cast on Mrs Bargrave's reliability. She was said to have seen ghosts before. Gray discovered that on one of these occasions her husband had gone to an inn in the country where he stayed for several nights with a whore. When Mrs Bargrave heard where he was, she went to collect him. Being told that he was in the garden, she went round to the back, whereupon the whore disappeared over the wall. When Mrs Bargrave mentioned that she had seen something like a ghost slipping over the wall her husband was delighted with 'so pretty a delusion to conceal his knavery', and afterwards used it to discredit her, saying that she was always seeing ghosts.

Gray solemnly recorded this tale and gave it as his opinion that 'the

Arguments for the truth of it are of much greater validity than those against it' (15 November 1705). This is the only place in Gray's correspondence where his religious values are clearly visible. He found Mrs Bargrave's membership of the Church and taking of sacraments as points in her favour. This might be expected, but the cool appraisal of the apparition followed by a statement in its favour is more surprising. The whole account brings clearly to mind that old women were still liable to be persecuted as witches in country areas. Mrs Bargrave's nervousness and caution in telling her story become much clearer when that chilling background is taken into account.

On the whole, belief in witches was becoming less common by 1700, although old women were still being tried in Scotland. Elspeth McEwen was tried and burnt in Kirkcudbright, accused of having a pin hidden in her rafters which she used to draw milk from her neighbours' cows, and of changing from the shape of a hare back to her own shape. This is one of the last recorded burnings and, significantly, the executioner had to be kept in prison beforehand to make sure that he would do his duty. In 1709, Elspeth Rule was found guilty of witchcraft, but was just burnt on the cheek. From 1700 onwards, a convicted witch was not likely to be burnt, and the most usual punishment was banishment.

Belief in witchcraft might have been repudiated by many scientists of the time, but their studies often show a continuing interest in what was strange. In his youth, Flamsteed had spent 'some part of my time in astrological studies' and, although he came to the conclusion that astrology gives 'generally strong conjectural hints, not perfect declarations' (Baily 1835), he remained interested in the subject. Among his papers is a copy of Thomas Hecker's *Ephemeris*, or astronomical table, taken from his astrological almanac for 1674.

Flamsteed wrote a preface, in which he was at pains to point out that the tables would be useful to astronomers even though predictions of the weather and affairs of state are 'pernicious' and an 'offence to Christian eyes'. Flamsteed then set out to prove from his studies and experience that predictive astrology is groundless. First, there were inconsistencies and errors in the way that the time of nativity had been assessed by such astrologers as Ptolemy and Kepler. Second, Flamsteed took the natal days of some of his acquaintances and of some famous men of the day and compared the predictions that their nativities required with the way these men actually lived. He gives as an example Bishop Hall, a 'learned divine', at whose nativity the Sun and Mercury were both cadent—causing any astrologer worth his salt to predict that the child born then would grow up to be more or less an idiot.

Flamsteed found such experiments sufficient to convince him that there are no satisfactory grounds for predictive astrology. Moreover, astrologers themselves, such as Gadbury and Lilly, could not agree on such fundamentals as how to treat the new heliocentric astronomy that Flamsteed and his colleagues considered to be the sound basis of all new work in astronomy. The telescope, the most wonderful instrument of the age, had shown that the Moon is a solid body with imperfections on its surface. The telescope would lead to further discoveries. Astrology was becoming merely irrelevant.

Nevertheless, the tradition was still strong. Flamsteed himself cast the horoscope of the new observatory in 1676 'for a laugh' for his friends. He also recorded his dreams, including one in which he met the King of France, who silently nodded to him but would not speak. This dream is inserted between a section on the deceitfulness of Isaac Newton and day to day notes on the activities of other scientists. Flamsteed would probably have subscribed to the view of dreams accepted from antiquity: they foretell the future.

Halley rejected astrology more firmly than Flamsteed. He made clear that the objections were not merely academic. The astrologer Gadbury was

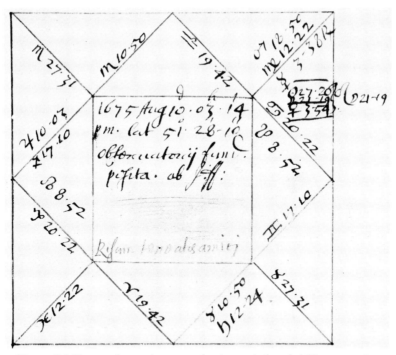

Figure 8.1 Flamsteed cast a horoscope for the newly founded Observatory in 1675. The twelve triangles round the central square represent the twelve mundane houses and the presence of the zodiacal constellations and the planets is noted in each one. At the bottom Flamsteed himself later added 'You may take this as a joke, friends.' Royal Greenwich Observatory.

suspected of being a Catholic and had ventured into political predictions, both of which made him dangerous to know or even mention. When Aubrey suggested to Halley that he should study astrology at Oxford, the reply was clear: 'As to the advice you give me to study Astrology, I profess it seems a very ill time for it when the Arch conjuror Gadbury is in prospect of being hanged for it' (16 November 1679).

Flamsteed, all his life, was a devout Christian. Many of his notebooks contain dedication of the work to God. Beyond this, Flamsteed had little trouble with his conscience or his religious belief as far as his writings show. He concerned himself with measurement, and all his work was compatible with the Protestantism of the day. Thomas Spratt, in his *History of the Royal Society*, had gone to great lengths to ensure that experimental science should be seen walking hand in hand with the Church of England. He depicted the experimenter as a 'doubtful, scrupulous, diligent observer of nature' who was thus inevitably also 'an humble Christian'.

Halley did not find the two strands of thought easy to bring together. His speculations about geology, and the Flood, in particular, brought his Christianity into question. In 1691, he applied for the position of Savilian Professor of Astronomy at Oxford, but was accused of believing that the world had always existed. After he had made clear that his theory of the Flood did not involve denying a divine act of creation and that this accusation was false, he was interviewed by Richard Bentley, later Master of Trinity College, Cambridge, on the more serious accusation that he had mocked religion and was an atheist. Halley protested that he was a Christian and expected to be treated as such. He never managed to dispel all doubts. Flamsteed was strongly against him, for various reasons, but irreligion was one of the overt ones. The remaining doubts against Halley were sufficient to keep him from the Professorship in 1691, but his atheism, if such it was, must have been very private. He remained friendly with Newton, who was himself deeply religious and would not have tolerated an atheistical friend. In fact, Newton himself was interested in the scientific and rational interpretation of the Bible. Richard Bentley used Newton's physics to argue for the existence of God, an endeavour that Newton entirely approved. He would have seen Halley's physical explanation of the Flood being caused by the near approach of a comet as rational and by no means irreverent. In 1703, Halley was sufficiently well respected to be given the Savilian Professorship in spite of Flamsteed's grumbling, and in spite of the requirement that still existed that all university teachers should subscribe to the Articles of Religion in the Book of Common Prayer.

Newton, who appeared so modern in his attitude to experiment and who set up the framework for subsequent thinking about the mechanics of the solar system, was still troubled by concepts of the seventeenth century. He objected most strenuously to accusations, particularly from foreigners, that his universal force, gravity, was an occult quality. He complained that Leibnitz 'changes the signification of the words Miracles and occult qualities that he may use them in railing at universal gravity'. Newton's gravity was a logical necessity and, therefore, in no way an occult quality, except in the sense that its cause is hidden from us. 'Miracles', in the sense of suspension of the laws of nature, play no part in the *Principia.* Yet Newton devoted much time and energy, particularly in the 1670s, to an understanding of the meaning of the Book of Revelation. 'God forecast everything of great moment', he wrote, and therefore it was a worthy task for him to study Jewish religion and prophecy to a point where he could understand what God intended. His manuscripts containing his conclusions extend to many thousands of words.

The Christian religion was as natural as breathing to astronomers of the seventeenth century but, by the beginning of the eighteenth, it was becoming less and less relevant to scientific work. William Derham is one of the exceptions. He was a priest of the Church of England. In 1714 he published an eight-volume work called *Astro-theology, or a Demonstration of the Being and Attributes of God from a Survey of the Heavens.* This work, together with *Physico-Theology*, was highly popular. It ran into nine editions, was translated into German, and was used as a source of material for eighteenth-century divines who wished to argue for the existence of God on the grounds that the existence of a watch demands the existence of a watchmaker. It was an expression of the biological consequence of the cosmology suggested by Newton's *Principia* (Meadows 1969).

William Whiston, Newton's successor as Lucasian Professor at Cambridge, also tried to combine religion with science. He had some success with his *New Theory of the Earth.* He tried to apply some elements of Newton's inductive method to an explanation of the origin and structure of the Earth. His theory said that the Flood described in Genesis had been caused by collision of the Earth with a comet, thus bringing together up-to-the-minute work on comets with his belief in the literal truth of Genesis. His religious fervour brought his career to an end when he let it be known that he no longer believed in the Trinity. Newton had reached the same conclusion, but had kept his belief quiet. Whiston was not greatly dismayed by his loss. He continued to preach, teach and write on his idiosyncratic views of religion and Judaic history. At the same time he made history by giving, in conjunction with Francis Hauksbee, a series of lectures on astronomy with demonstrations and experiments at Button's coffee house. The two facets of Whiston's activity are evidence that still in the eighteenth century not only could an intelligent use of experiment co-exist with a belief in the literal truth of prophecy, but that scientific method was seen as applicable to a study of revelation and religious belief.

The practical and observational tendency in astronomy is clear in the papers selected for publication in the *Philosophical Transactions* of the Royal Society from the early days. In 1670, the year in which John Flamsteed offered his first paper for publication, the astronomy papers as a whole were largely engaged in reporting practical observations:

> *Mr Mercator's considerations concerning the Geometrick and direct method of Signor Cassini for finding the Apoges, Excentricities and Anomalies of the Planets.*

> *A new planet appearing in the Swan Aug 17 1670 and the present appearance of the Planet Saturn observed by M Hevelius and others.*

> *The Moon's eclipse Sept 19 1670 The Conjunction of Venus and the Moon Oct 2 1670 with remarks on the new star in the Beak of the Swan and in the other in the Neck of the Whale M Hevelius etc.*

> *Mr Flamsteed's pre-advertisements of the Moon's motions this following year.*

Even Hevelius' comments on the new star are mostly details of its appearance and visibility, and are not concerned with theories of its nature or origin.

The other subjects of this volume of the *Philosophical Transactions* are similarly practical. The medical subjects sound strange to a modern reader because of the preoccupation with monsters and deformities, for example, 'The monstrous birth of a double child at Plymouth'. Yet this, too, is observation, in its own way, and much time was also spent in mapping out the normal anatomy, for example, 'A fuller discovery of the Vessels which convey the Chyle to the breasts of nursing women and that there is a suspicion of another passage of the urine to the Bladder than by the ureters'.

In medicine, as in mineralogy, physics, anthropology etc, a great deal of observing and collecting of oddities went on and was, perhaps, a necessary background to the development of a theoretical structure. In astronomy, theoretical developments were taking place on the Continent in the work of men like Cassini, Hevelius and Huygens, and the *Philosophical Transactions* in the 1660s and 1670s were dominated by their work on the solar system. Flamsteed's first paper marks the entrance of a new generation of British astronomers to the European scene. In 1676 the *Philosophical Transactions* still had papers by Cassini, Hevelius and Boulliau but there was also 'Mr Flamsteed's answer to Cassini', and, where 'Cassini remarks a huge spot in the sun', there is also 'Flamsteed and Halley on the same'.

The *Philosophical Transactions* continued to take on the function of bringing European work to the attention of the British, but in the decade of the publication of the *Principia* came increasingly the sense that there was work of great value to astronomy being done in England. The excitement of priority disputes took on added zest from patriotic fervour, and national pride became firmly established in the scientific scene. Thus, in 1687, the *Philosophical Transactions* published astronomical papers by Cassini, Eimmart, Wurtzelbauer, Haak, William Molyneux, Hevelius, Flamsteed, Wallis, de la Hire. In the 1690s, astronomy lost its prominent position in the *Philosophical Transactions*, and the signature on such papers as there were was almost exclusively that of Halley.

In 1698 an interesting change took place. The image of astronomy had been preserved until then as an independent science, and this was reflected in the use of a heading in the index. In 1698 astronomy was subsumed under 'mathematical discoveries'. Of these, only four are specifically in astronomy, but the mathematical connection makes clear that astronomy for the moment could be regarded more as a quantitative than a qualitative science.

Throughout the seventeenth century, mathematics had been a tool of astronomy. The publication of Newton's *Principia* in 1686 was a crowning achievement of the use of mathematics to describe and give precision to observable phenomena. Thus, the *Principia* are the mathematical principles that Newton was able to use to express the facts as they were observed. His thought could also move in the opposite direction, as his mathematical papers are often expressed in terms of the physics of motion. The *Principia* also had implications for cosmology. Newton's universe needed adjustments to preserve its stability and, here, both he and his followers saw a need for God.

After the *Principia* was published, there was a time lag before enough people understood its method and achievement to make its influence widespread. There were popularisers, such as the devoted Halley and, later, Voltaire, to transmit word of Newton's achievement to the curious of Europe. Once the method was widely known, it was understood that mathematics was used both to construct a model corresponding to the physical universe and then also to compare and contrast with the rules derived from experience, so that each could be used to modify the other. Moreover, the method could be used in other sciences, such as physiology, where it was used by Stephen Hales.

Some historians have suggested that the decrease in the number of papers and books on astronomy published in the years after the *Principia* until the 1720s or so was caused by a sobering awareness that an unattainably high standard of mathematical understanding and ability had been set by Newton and had left other practitioners feeling that they could not even begin. Others have suggested that there was a general feeling that Newton had finished all that there was to do in astronomy. Neither of these views of the period is borne out by a close look at the other astronomers of the time. Amateurs tended to ignore Newton's work because it had no immediate effect on their observations. They could continue to observe eclipses and occultations, unaffected by universal gravitation. On the other hand, Newton himself and his successors were fully aware of the gaps that Newton had left. He never achieved a satisfactory theory of the motion of the Moon, and the motion of Mars remained an unsatisfactory anomaly.

Probably the most influential part of Newton's work for his less gifted contemporaries was the emphasis that he placed on experiments which, in astronomy, are, in effect, observations. Newton's own statement of his method is clear: 'As in Mathematicks so in Natural Philosophy the investigation of difficult things by the Method of Analysis ought ever to precede the method of Composition. This Analysis consists in making Experiments and Observations and in drawing general Conclusions from them by Inductions and admitting of no Objections against the Conclusions but such as are taken from Experiments or other certain Truths.' Contemporaries saw in history a progression towards Newton's method. Roger Cotes wrote a preface to the *Principia* in which he pointed to three stages. First was the method of the ancients, which was philosophy with no experiments at all. Second came the mediaeval embellishment: hypotheses related to occult explanations. Third came Newton's method based on experiments and observations with no hypotheses. At the end of the seventeenth century Newton in this way had given shape and purpose to the method grounded at the beginning in Bacon's empiricism. The philosophers of the eighteenth century were to see themselves as applying Newton's method to all branches of knowledge.

Newton was the most famous and, therefore, the most influential advocate of experiment, but there were many others. Hooke not only tells that he regarded experiments as his approach to the truth but also sets up a standard for himself by which his principles must stand or fall: 'Nor wrest I any Experiment to make it quadrare with any Preconceiv'd Notion. But on the contrary I endeavour to be conversant in all kinds of Experiments and all and every one of these Trialls I make the standards or touchstones by which I try my former notions.'

The astronomers saw themselves as experimenters, and as specially bound to the truthful observation of nature. In this there was no antithesis to art or to literature. In the first half of the seventeenth century, scientific tracts and treatises were frequently introduced and interspersed with poetry. Jeremiah Horrocks, who observed the transit of Venus in 1638, included odes in his Latin account of his observation. Abraham Cowley wrote an ode to the Royal Society for Thomas Spratt's *History* in 1667. Halley still, in 1686, found it appropriate to prefix verses to the *Principia*. Members of the Royal Society included poets and writers like Cowley and Pepys, who should remind us that a dividing line between arts and science was just beginning to be drawn.

Although the visual arts were less well represented, the scientists themselves had considerable skills in drawing. Illustrations of experiments and observations were often detailed and usually gave a clear impression of the subject matter. Sometimes they were something more, extending to an interpretation rather than an exact literal representation.

Yet although the scientists still in many cases felt at home in the arts, the arts by the end of the seventeenth century were beginning to reject them. When writers took notice of the scientific virtuosi, it was usually to laugh at them. The ideas of the ingenious members of the Royal Society were written down in the *Philosophical Transactions* for all to see and many to mock. John Wilkins had popularised the idea that other planets, particularly the Moon, might support life, and that was seen by satirists as a hilarious example of speculation gone mad. Thomas Shadwell, the dramatist, made the hero of his play *The Virtuoso* an enthusiastic amateur scientist, Sir Nicholas Gimcrack. Sir Nicholas has a telescope, and with it he has examined the Moon. He has seen all sorts of detail there, including buildings and people fighting battles. They use gunpowder and those beasts of fantasy, elephants.

Elephants figured in Samuel Butler's satiric poem 'Elephant in the Moon' as well. Here the credulous scientist was again pilloried. The poem describes a group of virtuosi training a long refracting telescope on the Moon. Excitedly they claim to see a battle going on among the inhabitants, and one of them even sees an elephant on the surface. The credulity of those who believe all that they hear about what the telescope can show is mocked by the butler, the man from the street who brings common sense to the matter. He looks through the telescope but says that he cannot see these wonders on the Moon because something is blocking the tube. When the virtuosi eventually take the tube apart to examine it, they find their elephant—a mouse—and, further down, a litter of gnats and flies performing the rest of the battle.

Butler also wrote verses about the Royal Society, in which the rhyme and rhythm create a mocking effect, designed to increase public ridicule of pointless experiments:

> *These were their learned speculations*
> *And all their constant occupations*
> *To measure wind and weigh the air*
> *And turn a circle to a square.*

Such satires were undoubtedly popular. Shadwell's invention of Sir
Nicholas Gimcrack was an immediate success from the first production.
Robert Hooke went to see *The Virtuoso* in 1676. He felt very
uncomfortable as a representative of the Royal Society: 'people almost
pointed'. His opinion of Shadwell was not high: 'Mr Hill, Mr Hoskins and
I at Shadwell's atheisticall wicked play. 2½ sh. Cacht cold'. Mrs Aphra
Behn also had a popular success with *The Emperor in the Moon*, produced
in 1687. Here, a character pretends to be a man from the Moon in order to
seduce the daughters of a credulous philosopher. Some people, came the
message from the dramatists, will believe anything.

With the beginning of the new century, new satirists rose to fame, but
still the Royal Society and its scientists were a favourite target. Joseph
Addison and Richard Steele continued the mockery in *The Tatler* and *The
Spectator.* These two satirists found the collectors among the natural
historians worthy of particular notice. Although Robert Hooke had no
particular reason to feel attacked for this, as he was never a great collector,
some other astronomers were. Stephen Gray's interest in the bones found
at Chartham near Canterbury would probably have qualified for Addison
and Steele's attention. Addison paid Shadwell the great compliment of
taking up the character of Sir Nicholas Gimcrack again who, it seemed, had
recently died.

In his will Sir Nicholas had left his daughter Elizabeth a recipe for
preserving dead caterpillars, a preparation of winter maydew and an
embryo pickle. His younger daughter had three crocodile's eggs and his
humming bird's nest on the birth of the first child if she married with her
mother's consent. The oldest son was cut off with a cockle shell but the
younger, apparently the favourite, became sole heir to all the plants,
minerals, shells etc, including 'all the monsters both wet and dry' (Stimson
1949).

Collecting was not generally an activity of astronomers, and the majority
had neither the time nor the money to indulge in it on a large scale, even if
they had wished. The perpetual question of the layman: what is it all for?
would less often have been directed at astronomers than at others because
astrology accustomed the man in the street to think astronomy not only
useful but also inspiring, vaguely connected with religion, and not to be
questioned. For the more educated observer, astronomers were clearly at
work on the great practical problem of finding longitude at sea, effects of
refraction, variations of the compass etc. Tony Lumpkin, in *She Stoops to
Conquer* by Oliver Goldsmith, blunders about in the dark, saying that he
could 'as soon find out the longitude' as find his way. Richard Hogarth in
his series of engravings *The Rake's Progress* showed on the wall in the last
scene the 'mad astronomer' working on this same most intractable
problem.

Whether or not science was useful was the issue that aroused the interest of Jonathan Swift. When *Gulliver's Travels* appeared in 1726, the scientists of the Grand Academy of Lagado were immediately recognisable as the experimenters not only at the Royal Society but in workrooms and studies all over the country. In the Academy, scientists were trying to extract sunbeams from cucumbers so that they might be stored for future use. The impact of this idea on an age before the invention of solar heating panels must have been highly comic.

A number of experiments on the way in which plants breathe air, carried out by Hooke, Boyle, Grew and others, might have been the object of such mockery to those who knew about them. All sorts of work, some perfectly serious and worthwhile, might have been in the mind of a reader. Such satires show a low level of public understanding and sympathy for scientific work, and an acute need for better scientific education and information for the general public. An experiment may not be easy to justify in itself, but a whole project or series of experiments needs to be justified in terms of the main scientific programmes of the period, both for the sake of the scientists involved and for the benefit of the public.

As Newton was without question the outstanding figure of his age, it was in honouring him that the arts had the most enthusiastic contact with science. Newton's portrait was painted by an impressive list of painters, considering that the turn of the century was not a brilliant period in English painting. He was painted by Kneller, Thornhill, Vanderbank, Gandy and others. From the visual arts, the most satisfying tribute to science at the time may well be the statue of Newton that stands in Trinity College at Cambridge. As the value of Newton's work was appreciated by the French after Voltaire wrote, explaining and popularising what it meant, it is fitting that the sculptor was a Frenchman, Roubiliac (Haskell 1967).

Not all the poets and writers were satirical in their attitude towards science. Pope, one of the greatest English satirists, illustrates the change in attitudes that took place as the eighteenth century advanced. In his *Essay on Man* published in 1732–4 he showed his total allegiance to the world picture that was gradually emerging as a result of Newtonian science:

> *Nature and Nature's laws lay hid in night:*
> *God said 'Let Newton be' and all was light.*

Inevitably these attitudes fell before the assault of the Romantics such as Blake and Keats who, in their turn, complained that Newton and his admirers had tried to destroy the beauty of the rainbow by analysing it.

To the scientists in Newton's time, the satirists must have seemed to have the loudest voices. As a result, some may have felt persecuted and some may have modified their behaviour to be more in accord with current ideals of obvious usefulness. Yet, enough members of the Royal Society and of the scientific community continued undismayed for there to be a continuous output of work of all kinds and at all levels. Perhaps, at times, there was an uncritical acceptance of strange accounts and wonderful objects. Yet this same attitude allowed work to be admitted that turned out to be valuable in retrospect, but that any referee worth his post would have thrown out.

The atmosphere of curiosity that prevailed in the time of Newton allowed for the occasional exciting discovery in astronomy, but, in retrospect, the major achievements in the immediate shadow of the *Principia* were probably those of consolidation, filling in background, developing the instruments that would be used for new advances in the eighteenth century. For this kind of work, the amateur from any social level was able to play a part. Later in the eighteenth century, Maskelyne, the Astronomer Royal, achieved useful results, but it is William Herschel, the non-professional discoverer of Uranus, that the public remembers. Even though the importance of the professional was continually expanding, the fascination of astronomy has been such that amateurs have held on to a share of the activity. In spite of a much greater number of professionals and academics in the field, the first to discover a nova or an entirely new phenomenon could still be a schoolboy from Halifax—or a dyer from Canterbury.

Bibliography

Aubrey, John 1898 *Brief Lives* ed A Clark (Oxford: Clarendon Press)

Baily, Francis 1835 *An Account of the Rev. John Flamsteed* (reprinted London: Dawson 1966)

Boswell, James 1965 *Life of Johnson* (Oxford: Oxford University Press)

Capp, Bernard 1979 *Astrology and the Popular Press* (London: Faber)

Chambers, John 1820 *Biographical Illustrations of Worcestershire*

Clark, D H C and Murdin, L 1979 'The Enigma of Stephen Gray, Astronomer and Scientist' *Vistas in Astronomy* 23

Ellis, Aytoun 1956 *The Penny Universities: a History of the Coffee Houses* (London: Secker and Warburg)

Fleck, Ludovik 1979 *Genesis and Development of a Scientific Fact* ed R F Merton and T J Trenn (Chicago: Chicago University Press)

Frank, R 1973 'Science, Medicine and the Universities in Modern England' *History of Science* II

Haskell, Francis 1967 'The Apotheosis of Newton in Art' *The Texas Quarterly* 3 (Autumn)

Hill, Christopher 1965 *Intellectual Origins of the English Revolution* (Oxford: Oxford University Press)

Hunter, Michael 1962 *The Royal Society and its Fellows 1600–1700* (London: British Society for the History of Science)

—— 1981 *Science and Society in Restoration England* (Cambridge: Cambridge University Press)

Johnson, L W and Wolbarsht, M I 1979 'Mercury poisoning: A probable cause of Isaac Newton's physical and mental ills' *Notes and Records of the Royal Society* 34

Kearney, H 1970 *Scholars and Gentlemen: Universities and Society in Pre-Industrial Britain 1500–1700* (Ithaca: Cornell University Press)

King, Henry C 1955 *The History of the Telescope* (New York: Dover)

Koyré, Alexander 1978 *From the Closed World to the Infinite Universe* (New York: Harpers)

Macpike, Eugene 1937 *Correspondence and Papers of Edmond Halley* (London: Taylor and Francis)

Manuel, Frank E 1968 *Portrait of Isaac Newton* (Cambridge, Mass: Harvard University Press)

Meadows, A J 1969 *The High Firmament* (Leicester: Leicester University Press)

Parker, D 1975 *Familiar to All: William Lilly and Astrology in the Seventeenth Century* (London: Cape)

Porter, Roy 1982 *English Society in the Eighteenth Century* (Harmondsworth: Penguin)

Rigaud, S J 1965 *Correspondence of Scientific Men of the Seventeenth Century* ed J E Hofman (Hildesheim: Georg Olaus Verlagsbuchandlung)

Shapiro, B J 1968 *John Wilkins* (Los Angeles: University of California Press)

Spratt, Thomas 1702 *The History of the Royal Society in London* (London)

Stimson, Dorothy 1949 *Scientists and Amateurs* (London: Sigma Books)

Symonds, R W 1951 *Thomas Tompion, His Life and Work* (London: Batsford)

Taylor, E G R 1954 *The Mathematical Practitioners of Tudor and Stuart England* (Cambridge: Cambridge University Press)

Turnbull, H W (ed) 1961 *The Correspondence of Isaac Newton* vol III (Cambridge: Cambridge University Press)

Waller, Richard 1705 *The Posthumous Works of Robert Hooke* (London)

Westfall, R 1981 *Never at Rest* (Cambridge: Cambridge University Press)

Woodruff, C F and Cope, H J 1968 *Scolar Retia Cantuarensis* (London)

Notes on manuscript sources

Chapter 2

Flamsteed's autobiographical notes are to be found in the Flamsteed Papers at the Royal Greenwich Observatory with the reference numbers RGO 1/32, 1/40, 1/41, 1/48 and 1/62. A collection of letters from Flamsteed to Richard Towneley is held at the Royal Society in Mss 243 (F1). Correspondence between Flamsteed and James Pound is at the Royal Greenwich Observatory in RGO 1/37.

Letters from Stephen Gray to Flamsteed are in RGO 1/35 and others can be found at the Royal Society and in the Sloane Collection at the British Library.

Chapter 3

John Witty's correspondence is in the Flamsteed Collection in RGO 1/33 and 1/37. Abraham Sharp's many letters form the whole of RGO 1/34 and occur elsewhere in the Collection together with some of Flamsteed's replies.

Chapter 4

Correspondence between Flamsteed and Jonas Moore is in the Flamsteed Collection in RGO 1/33, 1/35, 1/36 and 1/37. The correspondence of Stanyan, Derham and Witty is mainly in RGO 1/37. There are many accounts of voyages and letters from sailors and naval personnel in RGO 1/41.

Chapter 5

Flamsteed's correspondence with his curates is scattered through the Flamsteed Collection at the Royal Greenwich Observatory. His accounts and financial records appear in the margins and are often mixed in with his observing notes.

The manuscripts for the Gresham lectures are in RGO 1/38.

Chapter 6

Letters from Brattle and Thomas to Flamsteed are in the Flamsteed Collection at the Royal Greenwich Observatory in RGO 1/37.

Flamsteed's correspondence with foreign observers, notably Hevelius, is in RGO 1/42 with scattered letters elsewhere.

Chapter 7

Descriptions of instruments are to be found in much of the correspondence mentioned already. Flamsteed's own account of his instruments is in his autobiographical notes (see notes to chapter 2). Some of Halley's and Bradley's instruments survive and can be seen at the Old Royal Observatory at Greenwich. Of Flamsteed's era only an angle clock by Tompion can still be seen at the Old Royal Observatory.

Chapter 8

Flamsteed's copy of Hecker's *Ephemeris* and his preface to it are at the Royal Greenwich Observatory in RGO 1/76.

Index

Anderson, Robert 134
Arbuthnot, Dr 135
Aristarchus 39
Aristotle 34, 38, 131
Ashe, George 91

Bacon, Francis 2, 3, 35, 142
Bainbridge, John 39
Barrow, Isaac 38, 74
Behn, Aphra 7, 144
Bentley, Richard 138
Blake, William 145
Bosseley, William 52
Boucher, Charles 61
Boyle, Robert 14, 145
Bradley, James 77, 114, 116
Brahe, Tycho 2, 5, 121, 124
Brattle, Thomas 28, 99, 100, 117
Briggs, Henry 39, 47
Brounker, William 85
Burnet, Thomas 30
Butler, Samuel 143

Cassegrain, M 109, 112
Cassini, Jean Dominique 9, 121,
  140, 141
Charles II, King 10, 12, 43, 44, 75,
  81, 89, 98
Cock(s), Christopher 85, 86, 111

Collins, John 28, 47, 48, 91
Copernicus, Nicholas 2, 4, 39
Cotes, Roger 20, 40, 41, 62, 82, 142
Cowley, Abraham 143
Crabtree, William 24
Crosthwait, Joseph 26, 62, 66, 116,
  126, 127, 128
Cutler, John 84

Dary, Michael 48, 74
Denton, Cuthbert 25, 68, 122, 123
Derham, William 6, 27, 64, 74,
  77–8, 116–18, 139
Desaguliers, John 56, 105
Descartes, René 24, 111

Flamsteed, John *passim*
Flamsteed, Margaret 64, 68, 100,
  126
Flamsteed, Stephen 13–18
Foster, Samuel 39

Gadbury, John 3, 14, 54, 137, 138
Galileo 2, 86
Gascoigne, William 24, 121, 122
George, Prince of Denmark 9, 95, 96
Godfrey, John 32, 71, 88, 105
Goldsmith, Oliver 144
Gray, Matthias 29–30
Gray, Stephen *passim*

150

Greatorex (Greatrackes), Valentine 14, 15
Gregory, David 23, 43, 57, 61, 82, 93, 95, 109
Gregory, James 20, 23, 43, 82, 111, 112
Grew, Nehemiah 145

Hadley, John 114
Hales, Stephen 141
Halley, Edmond 5, 17–19, 26, 28, 41, 49, 51, 57, 60–4, 72, 79, 83, 89–98, 107, 126–8, 132, 133, 137, 140, 141
Harris, John 88, 105
Hecker, Thomas 136
Herschel, William 146
Hevelius, Johannes 41, 64, 85, 114, 121, 124–5, 140, 141
Hodgson, James 26, 50, 68, 95, 117
Hogarth, Richard 144
Hooke, Robert 5, 10, 12, 20, 22, 30, 31, 41–9, 57–63, 70, 84–5, 107, 114, 119, 120, 125, 131, 144, 145
Horrocks, Jeremiah 24, 142
Hoskins, John 103
Hunt, Henry 30, 31, 58, 85, 102
Huygens, Christian 116, 140

Johnson, Samuel 82

Keats, John 145
Keil, John 23, 82
Kepler, Johannes 2, 39, 121, 136
King, Charles 39
Kneller, Godfrey 145

Leake, John 48, 49
Leigh, Luke 15, 16, 26, 27, 51, 52
Lilly, William 2, 137
Locke, John 59, 93

Marr, William 48
Marshall, John 111
Maskelyne, Nevil 146
Molyneux, Samuel 23, 43, 57, 127
Molyneux, William 83, 91, 121, 141
Moore, Jonas 10–13, 35, 44, 48, 49, 54, 75, 85–91, 107, 119–26, 133

Newton, Humphrey 59
Newton, Isaac *passim*
Newton, Samuel 50

Oldenburg, Henry 9, 58, 86, 87, 91, 111
Oughtred, William 22, 34, 43

Pagitt, Edward 49
Palmer, John 39
Paul III, Pope 4
Pell, John 47, 58, 63, 74, 88
Pepys, Samuel 49, 57, 64
Perkins, Peter 44, 49
Peter, Tsar of Russia 71
Petty, William 20, 22, 63
Pope, Alexander 133, 145
Pope, Walter 83
Pound, James 6, 23, 27, 74, 78–80, 114–16, 134
Ptolemy 4, 38, 136

Reeve(s), Richard 86, 111
Rooke, Laurence 23, 39
Ryley, Abraham 42, 51

Scarborough, Charles 56
Shadwell, Thomas 7, 143, 144
Sharp, Abraham 26, 27, 41, 62, 68, 71, 96, 106, 123, 126, 134
Shortgrave, Richard 118
Sloane, Hans 9, 30, 31, 68, 91, 95, 102, 135
Smith, Adam 35
Smith, Thomas 68
Spratt, Thomas 138, 143
Stafford, John 26, 123
Stanyan, Henry 66, 99, 100
Stephenson, Nicholas 52
Stileman, Timothy 76, 77
Street, Thomas 4, 54, 89
Swift, Jonathan 7, 145

Taylor, Brook 23
Thomas, Henry 99, 101
Thornton, Stephen 27, 74, 80, 117
Tompion, Thomas 58, 118, 119, 125, 126
Towneley, Christopher 24

Towneley, Richard 23, 87, 121, 122, 129

Vigani, John 59
Voltaire, François 141

Wallis, John 20, 35, 38, 61, 81, 82, 133, 141
Ward, Seth 38, 39, 43, 74, 80, 88
Weston, Thomas 68
Wharton, George 2, 23
Wheler, Granville 71, 105, 129, 135
Whiston, William 20, 39, 40, 43, 58, 101, 139

Wilkins, John 22, 25, 43, 74, 80, 133, 143
Wing, Vincent 4
Witty, John 51, 66
Wood, Anthony à 81
Woodward, John 30
Woolferman, Isaac 68
Wren, Christopher 5, 20, 22, 25, 41, 85, 107, 132
Wright, Matthew 27, 80, 117

Yarwell, John 111
Young, George 99